MEIJI RESTORATION LOSERS

HARVARD EAST ASIAN MONOGRAPHS 358

MEIJI RESTORATION LOSERS

Memory and Tokugawa

Supporters in Modern Japan

Michael Wert

Published by the Harvard University Asia Center
Distributed by Harvard University Press
Cambridge (Massachusetts) and London 2021

©2013, 2021 by The President and Fellows of Harvard College

Printed in the United States of America

The Harvard University Asia Center publishes a monograph series and, in coordination with the Fairbank Center for Chinese Studies, the Korea Institute, the Reischauer Institute of Japanese Studies, and other facilities and institutes, administers research projects designed to further scholarly understanding of China, Japan, Vietnam, Korea, and other Asian countries. The Center also sponsors projects addressing multidisciplinary and regional issues in Asia.

Library of Congress Cataloging-in-Publication Data

Wert, Michael.
    Meiji restoration losers : memory and Tokugawa supporters in modern Japan / Michael Wert.
        pages cm. — (Harvard East Asian monographs ; 358)
    Includes bibliographical references and index.
    Summary: "In this volume, Wert traces the shifting portrayals of Restoration losers and the supporters who promoted their legacy. By highlighting the overlooked sites of memory and legends, Wert illustrates how the process of commemoration and rehabilitation allows individuals a voice in the formation of national history"—Provided by publisher.
    ISBN 978-0-674-72670-3 (hardcover : alk. paper)
    ISBN 978-0674-25123-6 (pbk : alk. paper)
    1. Japan—History—Tokugawa period, 1600–1868—Historiography.
    2. Historiography—Japan—History—20th century.   3. Oguri, Tadamasa, 1827–1868.   4. Ii, Naosuke, 1815–1860.   5. Collective memory–Japan.   I. Title.
    DS881.4.W47 2013
    952'.025072—dc23

                                                                        2013009020

Index by the author
First paperback edition 2021

♾ Printed on acid-free paper

Last figure below indicates year of this printing

30   29   28   27   26   25   24   23   22   21

To my parents,
William N. Wert and Gwynn W. Wert

Acknowledgments

Many thanks to my mentors, first among them Anne Walthall, who taught me how to be a scholar. Kathy Ragsdale and Eugene Park read earlier versions of this project and provided helpful feedback, and Kathy patiently edited the entire manuscript. Kären Wigen has given much valuable advice and support over the past decade. Bin Wong and Ken Pomeranz contributed greatly to my training at UC Irvine, especially Bin, who always put me on the spot (for which I am grateful). I also thank Leo Hanami and the Asian Studies faculty at the George Washington University for helping me get started and teaching me Japanese.

Many scholars have given their valuable time, encouragement, and advice at various stages of this project, in no particular order: Douglas Howland, Morgan Pitelka, Mark Ericson (my Oguri *senpai*), Steven Ericson, James Huffman, Jordan Sand, Sarah Thal, Louise Young, and Julia Azari. Without HUAC editors William Hammell and Bob Graham, this project would have never seen the light of day. I thank them, two anonymous readers of the manuscript, and the publication staff at the Harvard University Asia Center and Westchester Book Services for their hard work.

I am grateful to Monbusho for funding to do research in Tokyo and to Marquette University for summer funding. At the University of Tokyo, Fujita Satoru kindly hosted my stay, and Yoshida Nobuyuki helped improve my *komonjo* and *kuzushiji* reading skills, which I started under Umezawa Fumiko and Kurushima Hiroshi. Yokoyama Yoshinori at the University of Tokyo's Historiographical Institute provided much assistance during my time in Japan, and helped me acquire the cover art. Like many Japan scholars of my generation, my journey began innocently enough in the Japanese countryside as a participant in the JET program in Gunma Prefecture. From there, serendipity led me into academia.

I thank the many kind people of Gunma, Kurabuchi, and Takasaki—in particular, Murakami Taiken and his family at Tōzenji; Kimura Naomi; Tōji Satoru and my Maniwa people; Suzuki Yasuhiro; William Jensen; Stacy Clause; and the Kojima family.

There have been many who made the project much easier with their encouragement and good humor. My graduate cohort at UCI, former colleagues at Oberlin and current ones at Marquette, students past and present, and many young scholars who have shared their knowledge and inspiration over the years: David Cannell, Michele Mason, Rod Wilson, Rob Stolz, Al Park, David Eason, Amy Stanley, Jeff Alexander, Franco Trivigno, François Blanciak, and Pieter de Ganon, among many others. Thank you to the staff at Stone Creek for keeping the coffee coming.

Last but not least, I must thank my family—especially my wife, Yuko, a life partner who makes every day better than the one before, and my baby goddess, Hera, who makes every day like Christmas.

Contents

Figures and Maps

MEIJI RESTORATION LOSERS

Introduction
Remembering Losers

This is a book about losers—men typically associated with the losing side of the Meiji Restoration. Most retainers who served the defeated Tokugawa shogunate during its final years survived the fighting, and some even worked for the new Meiji government in positions that made use of their expertise in military, diplomatic, or financial affairs. Some of the more famous ex-retainers remained active in a broad range of intellectually demanding endeavors: journalism, education, academia, and politics. But even as tensions between ex-Tokugawa men and the new government eased, the dominant narrative of the Restoration was not kind to the losing side. The Meiji-era proverb *kateba kangun, makereba zokugun* reflects this dichotomy quite nicely. Roughly the equivalent of "might makes right" or "victor's justice," it literally translates as "win and you are the emperor's army, lose and you are the rebel army."[1] Ex-retainers and their supporters did not accept the characterization lightly but expressed pride in their heritage through journals, biographies, apologetic histories, and commemorations.

Among the Restoration losers, I am most interested in those who suffered, to borrow Gary Fine's phrase, the most "difficult reputations": Oguri Tadamasa, Ii Naosuke, and, to a lesser extent, the Shinsengumi and the Aizu samurai.[2] The tortured posthumous history of Oguri, a Tokugawa bureaucrat who led the call to fight the Satchō forces, provides the narrative thread to my account. In keeping with the focus of college textbooks on the basic narratives of Japanese history, James McClain's recent textbook summarizes Oguri's notoriety succinctly: he "was the only shogunal official to be executed after the restoration."[3] The death of Oguri—killed in the countryside of Gunma Prefecture under suspicion that he would rebel against the Meiji government—shocked his contemporaries and informed all subsequent appropriations of him. Oguri is often compared to the more famously vilified figure Ii Naosuke. Although Ii was assassinated in 1860, long before the Boshin War, throughout the nineteenth and early twentieth centuries he was blamed for many of the problems that led to the Meiji Restoration, especially by Meiji oligarchs whose friends and mentors Ii purged in the late 1850s. Ii's death emboldened many of the pro-emperor zealots causing trouble in Kyoto, so the shogunate founded the Shinsengumi, one among several pro-Tokugawa police groups (some would call them vigilantes) in Kyoto and Edo that remained loyal to the shogun even after his resignation in 1867. The Shinsengumi's reputation—in particular that of Captains Kondō Isami and Hijikata Toshizō—swung between two extremes, defined alternately as bloody assassins and Tokugawa loyalist heroes. They have become the focus of popular culture in recent decades, and thus I address their memory mostly in the latter chapters of the book. Finally, the Aizu samurai received perhaps the harshest treatment at the hands of the Meiji military, and residents from the Aizu area suffered long-standing discrimination. Although their long and complex history deserves its own monograph, I trace their legacy through several chapters as it relates to the growing network of memory activists, those self-appointed custodians of Restoration losers.[4]

Memory activists consist mostly of local people who choose to support historical figures believed to be important to local and national identity. By commemorating the vanquished, these memory activists, bent on rehabilitating their fallen heroes, make claims to plurality in the formation of national identity, memory, and history.[5] Far from being passive participants in the construction of Restoration history and memory, local memory activists helped shape national interpretations of history through their commemorative efforts. They did not achieve this on their own but interacted with a range of people who could access a national audience: former Tokugawa retainers, politicians, writers, journalists, manga artists, and film producers. Activists also cooperated with counterparts who tried to rescue other losers from villainy and obscurity but clashed with those who would appropriate a loser's heroic legacy for selfish ends. As John Gillis notes, "commemorative activity is by definition social and political, for it involves the coordination of individual and group memories, whose results may appear consensual when they are in fact the product of processes of intense contest, struggle, and, in some instances, annihilation."[6] In so doing, activists use memory to interrogate history by challenging multiple interpretations of past events and offering their own alternatives within a national whole.

Only recently have scholars begun to study the historical memory of the Meiji Restoration. There are no books in English on the topic, but Carol Gluck's pioneering essay outlines the twentieth-century Restoration booms.[7] Saigō Takamori, perhaps the most famous Restoration figure, has become the focus of several article-length studies of historical memory and historiography. Of these, Mark Ravina's analysis of legends surrounding Saigō's suicide is the most similar to my own project. These "Saigō studies" in English share much in common with parallel developments in the Japanese-language scholarship.[8]

In Japanese, there are a few recent works concerned with memory and the Meiji Restoration, although the language and methodology of memory studies is usually absent. Narita Ryūichi has traced representations of the Restoration years and the Meiji period within the works of Japan's most prolific postwar writer, Shiba Ryōtarō.[9] Narita illustrates Shiba's immense role in shaping public interpretations of Restoration events. From the early 1960s onward, his views have been disseminated through literature, theater, print media, television, and film to a degree that has no equivalent in the West. Although Shiba's views changed over time, he generally celebrated marginal figures, men of low rank, and commoners trying to find their way in the tides of history. Narita uses Shiba to question the putative neat split between history and literature that became the focus of much postwar intellectual debate. Shiba, Narita argues, provides the best example of an intellectual who blurred this line, and it is through Shiba's writing about the period of Japan's early modernization (*kindai*) that we see the overlap between the past and the present.

Miyazawa Seiichi's *The Reinvention of the Meiji Restoration* (*Meiji ishin no saisōzō*) is the first book that attempts to trace how the Restoration recurred as a theme in historiography, politics, commemorations, and popular culture. He argues that during times of crisis from 1868 to the 1940s, people looked to the Restoration as a model for action. Modern Japan's early history is not a straight line from the Restoration to the future but a spiral, a constant returning to the past for guidance—an observation also recognized by Paul Connerton regarding the French Revolution's role in modern Europe.[10] This explains the booms in Restoration-related popular culture, biography, fiction, and historical writings during moments when political activists called for a "second Restoration."

Narita's and Miyazawa's studies are similar to the dozens of historical-memory projects on the American Civil War, the Revolutionary War, and the French Revolution. Although the Meiji Restoration was not as deeply traumatic as these other events, it nonetheless acted as a transitional event—a mythic origin to which intellectuals, politicians, and artists return to find both inspiration and blame. The dominant narrative of the Restoration, that both sides fought in the name of the emperor and for the progress of the country, grew in the fertile environment of Meiji-period reconciliation, despite tension among scholars over the causes of the Restoration and over whether it was complete or even beneficial. The seemingly smooth narrative has been challenged from the margins by memory activists and their allies from 1868 to the present. If, as Gary Fine suggests, historical reputations are a shorthand for how we think about a time period or an event, then the struggle by memory activists to rehabilitate their forgotten or vilified heroes is a struggle over the interpretation of the Restoration itself.[11]

What is missing from all of the above-mentioned works is the role of local and marginal historical memories. The central argument in this book is that local commemorative efforts by memory activists have, over time, changed regional and, more significantly, national interpretations of the Meiji Restoration. I am not advocating a rescue of lost voices, nor celebrating a simple bottom-up view of historical interpretation. Instead, taking a cue from Alon Confino's work, I look at the middle, where the national and local interpretations of the Restoration meet, effecting change in both. Rather than focusing solely on the political questions that memory poses—Who does the remembering and why? What do the conflicting memories say about rival political groups?—scholars need to look to places where memory is unexpected—in the private sphere of personal connections, in campaigns to boost tourism, in popular culture—to see how memory looks different outside the sphere of political activity. By connecting seemingly unrelated topics, we can discover the common denominators that define a shared sense of national identity despite political differences.[12] "Losers" are particularly useful in achieving this. By definition, they are on the margins of history, where rehabilitating them requires more work than does lauding well-established heroes.

A central feature of memory work is conducted around what Pierre Nora has termed "sites of memory" (*lieux de mémoire*). Nora edited a multivolume study of well-known objects, symbols, songs, and museums that connected the French citizenry to the nation. Although the French version includes a few regional and local sites of memory, he has nonetheless been taken to task for being top-heavy in his analysis.[13] As is often the case with scholarly concepts, scholars in other disciplines and fields have appropriated the concept "sites of memory" in their own ways. In Japanese studies, Takashi Fujitani demonstrates how these sites created the modern Japanese nation-state, a subtle but important departure from Nora, who assumed that the French nation existed prior to memory. Another adaptation has been to use "sites of memory" to refer to lesser-known local monuments, statues, and so on.[14] These micro sites also contribute to a definition of the nation, even if they are not well known or seem to challenge dominant historical narratives. These local sites become particularly powerful when they combine to form a memory landscape. Memory landscapes are created from the network of sites where memory takes place. They narrate a historical individual's legacy through physical markers where commemoration is explicit, such as monuments, graves, and statues, and, in places where memory seems more fuzzy and inconsequential, through objects and events that are otherwise mundane.

Memory landscapes become powerful matrices when activists consciously work to situate their heroes in a larger "landscape." A lecture given in Aizu-Wakamatsu City about Oguri by a local researcher from Gunma, for example, is not coincidental but represents an effort to connect with similarly marginalized "loser" histories, such as the Aizu legacy. The football field–sized hole scarring the side of Mount Akagi, and the Akagi golf course—putative locations of Tokugawa buried treasure—became sites for obsessive gold digging for adventurers as well as television producers around Japan. It is in these ways, through the explicit and the implicit, that reputations sustain collective memory by finding an audience with whom the subject shares common values and beliefs.[15]

Thus, the concept of "memory landscapes" forces commemoration to be viewed as a *process*. Here I do not mean "process" in the way it often does in memory studies: the simple back-and-forth creations and appropriations of discourse. Instead, memory landscapes highlight the tangible means required to, for example, convince NHK to produce a television show about a historical figure, or a city to donate space for a statue. This helps answer the tricky question of reception. It is one thing for a memory activist to evoke an agreeing nod from a passive listener; it is an entirely different challenge to coax someone into devoting time and money to a cause. The latter requires a powerful message that plays on the symbols and values that dominate any historical moment.

Memory landscapes are historical and thus change over time. Although today Oguri might be associated with Gunma Prefecture, this was not always the case. During the 1930s and 1940s, the Fumon'in Temple in

Ōmiya City, Saitama Prefecture, received national coverage as the premier site of Oguri's memory following an article and visits by several Japanese celebrities (see chapter 3). Shortly after Prime Minister Okada Keisuke visited Oguri's grave in 1935, the novelist and politician Itō Chiyū remarked, "Now that Oguri's image has been rectified, there is nothing left to rewrite about Meiji Restoration history."[16] Figure 1 illustrates Oguri's memory landscape as it existed in 1935, as depicted in Nakamura Kaoru's *A Cultural History of Kanda* (*Kanda bunkashi*). Even the book itself is part of the memory landscape—his section is titled "Oguri Kōzukenosuke, the Great Man of Surugadai," the area where he was born and raised.

Except among diehard fans, Oguri would hardly earn such accolades in Japan today. But these statements raise two issues: why was Oguri deemed so important, and why did the degree of his importance fluctuate over time?

Thus, the first chapter presents a case for why Oguri became a unique figure for commemoration by a broad range of people. What was it about Oguri, for example, that prompted the novelist Nakazato Kaizan, who featured Oguri in parts of his *Great Bodhisattva Pass* (*Daibosatsu tōge*), to state in no uncertain terms, "there is no figure as important as him in the history of the Meiji Restoration years." Although the main goal of this brief biography is to establish the basic facts of Oguri's life and death as best as can be understood from the few written sources and oral histories available to historians, the narrative itself serves to create an impression of how his contemporaries later appropriated his legacy. Their writings about Oguri formed the foundation that subsequent memory activists had to address. Thus, I begin with Oguri's beheading, an event that would later highlight his portrayal as a martyr. I then move on to his controversial policies, his interaction with commoners, and his final weeks on his lands in Gonda Village. Finally, I briefly outline the local aftermath of his execution.

Journalism after 1868 was dominated by former Tokugawa men who wrote counter-narratives of the Restoration to qualify the official version advocated by the Meiji oligarchy. The second chapter outlines the initial historical memory of Oguri, Ii, and Tokugawa retainers within the larger context of history production in the Meiji period. It focuses on textually mediated historical memory: the stock of stories about Oguri, and how those who appropriated him negotiated the changing face of the memory landscape. Former Tokugawa retainers attempted to rescue Oguri's story from a dominant historical narrative that marginalized and vilified him and, in so doing, hoped to protect their own legacy. They used his story

巨人小栗上野介の遺跡

上州権田村烏川畔の碑「偉人小栗上野介・罪なくして此所に斬らる」　小栗上野介の胸像（横須賀公園）

小栗上野介の塚（権田村東善寺）

大宮町・普門院境内の小栗上野介招魂碑

上州烏川畔・上野介の斬殺せられし河原

Fig. 1. Memory landscape. From top to bottom, left to right: Oguri's mound, Gonda Village; Oguri memorial, Fumon'in Temple, Omiya; Karasu River, landscape where Oguri was killed; monument where Oguri was killed, Karasu River; Oguri's bust, Yokosuka. (Nakamura, *Kanda bunkashi*, 28)

to attack the Meiji oligarchy, highlighting its unjust actions during the Restoration, and to shame former colleagues who betrayed their heritage by working in the new government. In the Tokyo hinterland, where Oguri was killed, his legacy remained within the realm of communicative memory: everyday rumors, legends, and gossip absent from traditional historical sources—a memory that owed its longevity to the persistent buried treasure legend attached to Oguri's legacy.[17] The Tokugawa buried-treasure story is the most famous of its kind in Japan and surfaces during times of crisis: the late nineteenth century, the 1930s, the immediate postwar period, and the 1990s. Thus, I weave it throughout each chapter, dovetailing with more familiar modes of history and memory production.

In the third chapter, I seek to understand how memory activists envisioned alternative narratives of the Meiji Restoration by creating physical objects that anchored the past to the present. Debates about monuments often focus on the degree to which sites can be accurately "read" by visitors. This chapter demonstrates that accuracy in this context is a nonissue. Promoting an anniversary celebration or asking for donations to erect a statue produces discourse, adding to textually mediated historical memory that outlives an object's popularity or readability. I begin the chapter with Ii Naosuke supporters who tried to erect Ii's statue in a Tokyo park, only to be thwarted by the Meiji government on the grounds that Ii was still a villain. Fujitani has ably illustrated how the Meiji oligarchs transformed Tokyo into an administrative and symbolic center. The attempt to erect Ii's statue forced the oligarchs to create laws limiting the use of public space in the capital, demonstrating that even national space has to respond to marginal voices. Although they failed in Tokyo, memory activists succeeded in erecting a statue of Ii in Yokohama during the city's fiftieth anniversary (1909) despite protests from several Meiji oligarchs.

I compare Ii's case to Oguri's commemoration at Yokosuka City, a key naval port in Japan's modern naval victories, during its fiftieth anniversary (1915). Supporters of controversial figures such as Ogyū Sōrai, Ii, Oguri, Sagara Sōzō, and the Aizu samurai also attempted to obtain posthumous court rank for their heroes. A renewed phenomenon during imperial Japan, the value of posthumous court rank was determined by the recipient's proximity to the emperor. Local memory activists with access to historical resources and government connections could recommend potential awardees, and often the Imperial Household Ministry relied on them to do so. The different modes of commemoration exemplified through statue building or lobbying for posthumous court rank led to Oguri's emergence as a regional hero. But far from encouraging harmony at the local level, villages and temples argued over the location of Oguri's decapitated head and, thus, his legacy, while treasure hunters clashed with Oguri's descendants and supporters.

The fourth chapter addresses how Japanese grappled with the WWII experience through the lens of postwar popular culture centered on Restoration losers, and the postwar boom in local memory activism. Novelist Ibuse Masuji's *Priest of the Fumon'in* (*Fumon'in san*), a short story about Oguri's executioner, and Shochiku's thirty-fifth anniversary period film *The Birth of Tokyo, the Bell of Oedo* (*Dai Tokyo tanjō, Edo no kane*) attempted to deal with wartime trauma by reimagining Oguri's martyrdom. Ultimately, Oguri failed to become a national hero, while other Restoration figures who were previously ignored gained national attention. The 1970s "old hometown" (*furusato*) boom recalibrated the relationship between the national and the local, giving more opportunities for prefectures to create and promote local identity. Mirroring trends elsewhere, people created local Oguri study groups and formed partnerships with other regions connected to Oguri, such as a sister-city relationship between a village in Gunma Prefecture and Yokosuka that celebrated business connections as much as it did a shared identity.

Scandals, natural disasters, and the crisis of identity that characterized the "lost decade" of the 1990s ripped apart the old myths valorizing the Meiji legacy. Distrust of the political, economic, and cultural status quo changed contemporary historical consciousness about the Meiji Restoration period, allowing past losers such as Oguri and the Aizu fallen to reappear on the national stage. WWII dominated the memory-discourse boom experienced during the 1990s, but local voices were usually crowded out by national ones. Local citizen groups appropriated the Boshin War anniversary to participate in the national conversation about memory. Likewise, an "Oguri campaign" taken up by Gunma Prefecture resulted in national television coverage for Oguri and his eventual inclusion in school textbooks—a final rehabilitation. The last chapter demonstrates how Oguri memory activists solidified their networks with Tokugawa family descendants, professional historians, and memory activists elsewhere who sought rehabilitation for other martyred men. This local-national-global interaction is a testament to the malleability of commemoration, memory, and historical production not just by national figures who influence mass media and culture but also by common citizens who have a stake in defining national identity.

CHAPTER 1

The Last Bannerman

On the morning of the sixth day of the fourth intercalary month in 1868, imperial troops escorted Oguri Tadamasa from his temporary imprisonment to the banks of the Karasu River in Mizunuma Village. Typically, capital punishment for a high-ranking samurai, especially a direct vassal of the shogun, involved a ritual gesture of suicide before an executioner's coup de grace. That day, however, Oguri was forced to bend over, hands tied behind his back. Besides calling the man who dared push his body forward with his feet a "disrespectful lout," Oguri's final words were a request to let his wife, daughter-in-law, and mother go free. A low-ranking samurai struck Oguri's neck not once but three times before his head dropped unceremoniously into a pit. A villager who witnessed the execution as a boy recalled, "What was most impressive in my mind was how white the soles of his *tabi* appeared when the body fell forward."[1]

This scene weighs on Oguri's commemoration, coloring explanations of his career up to that moment. It marks the origin of his commemoration both geographically, as ground zero for the historical memory about him, and temporally—almost immediately after his execution, former colleagues became his first apologists, protecting his legacy in death though they could not help him in life. The goal of this chapter is to impart a historical understanding of Oguri and clarify why memory activists and supporters found him a compelling figure worthy of appropriation.

My account necessarily begins with an outline of the Oguri family itself. Not all bannermen retainers were created equal, and understanding the Oguri family's connection to the Tokugawa clan and the shogunate provides a sense of Tadamasa's stature even before his first major career appointment. His family's history, moreover, affected his relationship with fief villagers, most importantly those in Takahashi Village and Gonda Village. Next, I move to a brief description of his official duties, focusing on his major contributions to the shogunate and Japan, the controversies with which he was involved, and the challenges he faced. Finally, I cover his brief life in Gonda Village, including the disorder and political climate in the region, his execution, and his family's escape from persecution.

History and memory intertwine in the primary source material for this chapter. Most primary documents relating to Oguri's official work can be found in the *Manuscript of Historical Materials on the Restoration (Dai Nihon ishin shiryō kōhon)* in the Tokyo University Historiographical Institute. The latter half of this chapter, however, concerns Oguri's personal relationships and interactions with people living in the countryside. Here I draw on extant portions of his diaries and on local history. His diary and account books were discovered accidentally in 1956, and their history reflects his violent death—confiscated by the Tōsandō army, inherited by the first governor of Iwahana Prefecture, found by a prefecture functionary named Hachirōemon, then hidden and forgotten in his house until discovered by his grandson eighty years later.[2] Like many diaries written by men of Oguri's rank, there is precious little commentary, much of it recording his daily routines from 1/1868 until four days before his execution: appointments kept, visitors received, and money spent. The longest and most detailed entries cover the last months of his life, from the time he arrived in Gonda to his arrest. Sources for local history include village, city, and prefecture collections commonly used by professional scholars, as well as oral histories collected by local historians. I use their work with caution, as historians are also agents of commemoration, but to discount their work because of this creates a false dichotomy between history and memory that this book seeks to dispel.[3]

The Oguri

The Oguri possessed an impeccable pedigree for a Tokugawa retainer family, due to their proximity to the Tokugawa progenitor, Ieyasu.[4] The given name for all Oguri male heirs, Mataichi, signified their high status within the Tokugawa retainer community. Not only did the name Mataichi indicate a unique relationship between the Oguri and the Tokugawa shogunate, but it invoked martial valor. Both the third- and fourth-generation Oguri men served Tokugawa Ieyasu. The fourth-generation Oguri, Shojirō, fought alongside Ieyasu during the Battle of Anegawa (1570). In what became the defining event for Oguri identity, Shojirō saved Ieyasu's life by grabbing a spear and killing an enemy who had suddenly attacked Ieyasu, catching him off guard. Ieyasu gave Shojirō the spear as a reward, thus raising the young warrior's profile among Ieyasu's retainers. Later, during the battles of Mikatagahara (1573) and Nagashino (1575), other samurai were said to have commented, "Yet again Oguri is the best with the spear" or "Will Tadamasa again be the best with the spear?" Ieyasu subsequently awarded Oguri the name Mataichi ("again, number one"), and both the name and its legacy have passed to each Oguri male heir, including the current one—a manga artist.[5] During one of the last battles against Ieyasu's enemies, the summer campaign at Osaka Castle, Shojirō was hit in the stomach by an enemy bullet and died of his injury in 1616, only five months after the death of his lord Ieyasu. Shojirō's remains were buried on the grounds of the Fumon'in Temple (Saitama Prefecture), where a monk would later become a contender for the Oguri legacy in the mid-twentieth century.[6]

Although not among the highest-ranked retainers of Ieyasu, the Oguri had the distinction of being assigned fiefs as a source of income. Among the roughly seventeen thousand Tokugawa vassals, only the upper echelon retainers known as bannermen (*hatamoto*) drew income from fiefs, and by the nineteenth century, fewer than half of the five thousand bannermen commanded fief income.[7] Of these, the average bannerman received between 500 and 600 *koku*. According to one bannerman study, only 36 percent ranked above 1,000 *koku*, and fewer than half received income from fief lands, mostly located in the Kantō region surrounding Edo.[8] Like those of other bannermen, the Oguri fiefs were spread widely throughout the Edo hinterland, totaling 2,500 *koku*. After Tadamasa's return from the United States in 1860, he was awarded another 200 *koku*. The oldest and largest of the fief's eleven villages was Takahashi Village, assigned to the Oguri by the shogunate in 1661.[9] The three largest villages after Takahashi were all located in present-day Gunma Prefecture, with others in Tochigi, Saitama, and Chiba prefectures.

Despite the long history between the Oguri and their fiefs, little is known of their interactions until the late eighteenth century. Most bannermen never visited their fiefs; the only contact between a bannerman and a village might be the occasional visit from a village representative, or if there was unrest, a bannerman might send his vassals to deal with the problem. Moreover, fiefs could be confiscated and reassigned by the shogunate at will, leaving little incentive for either the bannermen or the fief villagers to deepen their relationship with each other. Still, some

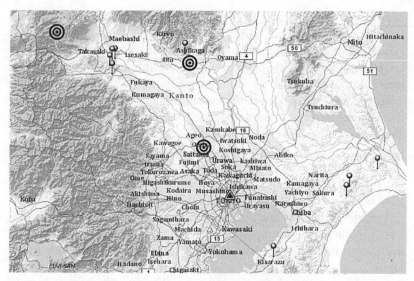

Map 1. Oguri Tadamasa's fief villages, Fumon'in, and Edo residence.

bannermen used servants in ceremonial visits to the court at Kyoto, or as foot soldiers (*ashigaru*) during processions at the Ise and Nikko shrines. Commoners of a lower status might be used to carry weapons, while village headmen—those given the privilege of using a surname and carrying a sword—filled more prestigious positions.[10] As with other bannermen, the Oguri walked a fine line with their fief representatives, drawing from them material and human resources while trying to avoid overexploiting them. If a bannerman demanded too much from his fiefs, peasants could refuse to provide goods and services, or even threaten to protest. Unlike daimyo lords, bannermen had little recourse if a peasant protest caused regional disorder. Worse, a bannerman could be punished for incompetence by having his fiefs confiscated. The Oguri suffered such problems with Takahashi Village, making it a less attractive post-Edo home for Oguri Tadamasa.

The late eighteenth-century rule of Takahashi by Tadamasa's grandfather was plagued by administrative problems, all of them arising from construction projects on the Watarase River. These projects had been particularly onerous for Takahashi Village and the surrounding area since the 1730s.[11] The first signs of corruption began in 1781, when peasants sent a petition to the finance commissioner requesting that the village headman, Sōbei, be fired.[12] Calling him selfish, they accused him of plotting with one of Oguri's retainers to rob the village of money.[13] In addition to assigning desirable jobs only to his friends, Sōbei forced villagers to work without pay on the river, and those who were paid received only IOUs. Although the shogunate funded the project, Sōbei informed peasants that the work payments would be taken out of the village budget. The incident reflected badly on the Oguri family's rule over the village; the villagers not only requested their headman's dismissal but also demanded an investigation of the Oguri retainer and village elites who benefited from the corruption. At one point, the entire village threatened to march on the retainer's office, but two local monks intervened and prevented the situation from escalating.[14] These events strained relations between the Oguri family and Takahashi Village for generations.

Despite the active Oguri commemoration in and around Gonda Village throughout the twentieth century, even less is known about its connections to the Oguri family until the 1860s. Having been added to the Oguri land portfolio in 1705, Gonda was the second largest village after Takahashi, and provided a steady source of lumber. Taxes were high in Gonda—so high in fact that when peasants in the neighboring village of Mizunuma received a new bannerman in 1837 and protested his 10 percent tax increase, which would have put the village on par with the 80 percent tax in Gonda—the bannerman quickly backed down.[15] Another incident, in 1854, reveals how little administrative contact there was between the Oguri and Gonda. When a local Gonda peasant absconded, the Edo city commissioner noted that Gonda lacked a fief office and depended on the local shogunate intendant at Iwahana to deal with investigations, including searching for the missing peasant. Inquiries and reports regarding the incident went back and forth between the Edo city commissioner, the finance commissioner, and village officials; the only involvement of the Oguri family consisted of a brief interview with a retainer in Edo.[16]

Nevertheless, there is evidence that the personal connections between the Oguri family and Gonda elites were deeper than those in other fiefs. For example, the sixth-generation Oguri Masashige renovated Gonda's only temple, Tōzenji, and contributed a memorial tablet with his father's and grandfather's names. The eighth-generation Oguri Nabeshirō lived in Gonda with a sake-brewing family, the Makino, probably convalescing in the healthier mountain air, and died in 1744 at age twenty-eight. A memorial stone with Nabeshirō's and Makino Chōbei's names still stands in the Makino family garden.[17] The Tōzenji possesses a document stating that the oldest son of the ninth Oguri had completed his training to become a monk, which indicates that he might have spent time at the Tōzenji before dying in 1781 at the age of twenty-nine. But the connection that set

Gonda's relations with the Oguri apart was the partnership between the village headman and Tadamasa's father, Tadataka.

Tadataka was not born into the Oguri family, but he saved them from certain extinction. A series of dying heirs endangered their existence. When the ninth-generation heir died, his younger brother took over the family, but he died young, as did his estranged wife and four daughters, all of natural causes. Luckily, he had a son, Tadakiyo, but when Tadakiyo fell ill at the age of seventeen, the family took no chances. They soon adopted the youngest son of a neighboring bannerman family. He became Oguri Tadataka, the eleventh-generation heir who led the Oguri family until his death in 1855.[18]

Tadataka employed two men who were later instrumental in his son's life. The first was Satō Tōshichi, the Gonda Village headman who married into the Satō family on Tadataka's recommendation. He later became a trusted Oguri vassal, traveling with Tadamasa on the first Japanese embassy to the United States in 1860.[19] The other fateful employee, Kimura, was initially not as impressive. An illiterate youth from a dubiously low-ranking family, Kimura worked temporarily as a guard at the Oguri mansion in Edo and befriended young Tadamasa, who, born in 1827, was only a few years younger.[20] Kimura soon left his appointment with the Oguri to marry into a confectionary merchant family, the Kinokuniya (now a major bookstore chain), and changed his name to Minogawa Rihachi. In 1852, he saved enough cash to buy into a money exchange company.[21] He would later change his name to Minomura and run the Mitsui merchant group, navigating it through the Meiji Restoration by using his connections to Oguri Tadamasa.

Tadamasa grew up in an intellectual milieu typical of nineteenth-century samurai living in Edo. In 1814, the Oguri compound briefly hosted the neo-Confucian scholar Asaka Gonsai. It is unclear when he left, but he moved around Surugadai until he opened his own school in 1827. The shogunate eventually hired him to teach at its official academy, the Shōheikō, in 1851. Asaka's career had peaked during the growing threat of Western presence in East Asia. He became an advocate for strong coastal defenses and "expelling the barbarians" from Japan while avoiding war.[22] He also believed that Japan should create a merchant fleet, as commerce would one day strengthen the country.[23] During his career, Asaka taught over two thousand students, including key shogunal retainers, future Meiji intellectuals, and samurai from domains that opposed the shogunate.[24] Like Tadamasa, many of Asaka's students—such as Kimura Kaishū, Kurimoto Joun, and even non-shogunate men such as Yoshida Shōin and Takasugi Shinsaku—interacted with Westerners. Tadamasa did not adopt

Asaka's xenophobia, but he shared his sense that Japan needed a navy in order to advance its standing in world affairs.

A Controversial Career

There is little reliable information about Oguri before his first significant career appointment in 1859. He passed through a series of positions common for a bannerman of his status until 1859, when, for reasons unclear, Ii Naosuke selected him to serve as an inspector (*metsuke*). Ii promoted him in 11/1859, officially granting him the title Bungo no kami, his typical designation in the American press.[25]

Beginning with his accompaniment of the foreign commissioner to the French consulate on 10/11/1859 to negotiate compensation for a Chinese servant killed by rōnin, Oguri's entire subsequent career involved interacting with foreigners.[26] He was particularly celebrated for his role on the 1860 embassy, about which, unfortunately, he left no record. We know that in Oguri's case, the title of inspector was usefully vague enough to allow for a range of responsibilities; he was expected to work as an advisor and a replacement should either of the other ambassadors become incapacitated, and the scholarly literature typically includes Oguri as part of the triumvirate leadership.[27] His most notable contribution to the embassy, an event often cited by memory activists, involved his actions at the Philadelphia mint. Due to unfavorable coinage exchange rates, Western merchants had been buying gold coins from the Japanese and selling them abroad at a profit. The shogunate hoped to stop gold from draining out of Japan, and Oguri attempted to achieve this by demanding a reassessment of Japanese and American coins. The assayers acquiesced, though the new assessment failed to solve the monetary crisis.[28]

Oguri's first assignment to duties abroad had immediate consequences for his family and fief. The Oguri family experienced difficulty producing heirs, but Tadamasa's health was also a concern—a childhood disease had left his faced pockmarked. He wasted no time ensuring that the lineage would continue in case he died during the embassy. Only a day after his final promotion, he adopted a daughter named Yokiko from his paternal uncle, Kusaka Kazuma. Once Oguri received the ceremonial delegation on 12/1/1859, he found a husband for Yokiko, who married on 12/26.[29] Oguri's new son-in-law took the name Mataichi and became the heir to the Oguri family. Oguri also chose the Gonda Village headman, Satō Tōshichi, to accompany him on the embassy, making him, as locals like to claim, the first Japanese farmer to travel around the world.[30] Satō served in the official capacity of Oguri's retainer long after the embassy. In 1862, he served on Oguri's staff in Edo during Oguri's tenure as the Edo city commissioner, and in 1867, Oguri assigned him to work for the Japanese side of the Paris World Fair.

When Oguri returned to Japan in 1861, he advocated a shogunate-first approach to domestic rule, even though the political mood in the shogunate had shifted away from the hard line approach of Ii Naosuke, who had been assassinated, to a more conciliatory one with the court and daimyo. The embassy received a cool reception, but members still received promotions for their time abroad. It was for his service in the embassy that the shogunate granted Oguri an extra 200 *koku* in the form of two villages in Kōzuke Province. The shogunate then assigned him, as the new foreign commissioner, to convince a Russian naval captain to leave Tsushima Island. In the spring of 1861, captain Nikolai Birilev of the Russian corvette *Posadnik* had arrived in Tsushima claiming that his ship was too damaged to leave and had requested a meeting with the local daimyo in the hope of leasing land on Tsushima to give Russia access to a warm-water port. Oguri and Birilev bickered for days over the legality of Birilev's actions and his desire to meet with the local daimyo. Their final meeting on 5/18/1861 did not go well. Oguri tried to renege on a promise to allow Birilev to meet the local daimyo in a week's time. Birilev pressed Oguri on the issue, but Oguri claimed he had made such a promise without permission from Edo: "this is my responsibility and I will not agree [to your meeting the daimyo]. If you don't concur then you should shoot me as you see fit."[31] During the 1930s and 1940s, supporters used this statement as evidence of Oguri's steadfast samurai ethic.

Ultimately, the shogunate asked the British to scare off the Russians. Although this proved a successful tactic, in Oguri's mind, the Russian intrusion into Japanese waters had presented an opportunity to assert the shogunate's authority over strategically located domains, an opportunity that the shogunate had failed to exploit. Frustrated with the shogunate's orders to travel to Hakodate to negotiate with another Russian, and then another order sending him back to Tsushima, Oguri simply returned to Edo and asked for permission to resign.[32] The shogunate informed inspectors that Oguri's trip to Hakodate had been canceled "due to illness," perhaps to cover up dissent among its ranks.[33] Either way, Oguri and his assistant resigned their posts on 7/26/1861, thus ending Oguri's short and difficult career as a foreign commissioner.

In a letter to his superior, Mizuno Tadanori, dated 7/5/1861, Oguri complained about the difficulty of working with the Tsushima house elder and about how officials there ignored shogunate protocol. Moreover, he claimed the Tsushima officials never followed his advice.[34] This last complaint was one he had about not only the Tsushima officials but the shogunate as well. In another letter, Oguri stated the following:

> The year before last when I gave my opinion on raising the price of the *koban* coin, it was not followed at all. Although they [the shogunate leaders] agreed with my ideas on gold and silver coins last year after I returned from the embassy, they do not listen to me now. When I gave an opinion regarding foreign ministers returning to their countries, they did not heed my advice. I give advice with much seriousness, but it is never adopted.[35]

Oguri's strategy for solving the Tsushima problems differed fundamentally from that of many of his colleagues and seniors. Oguri, like Ii Naosuke before him, believed that the shogunate should take direct control of Tsushima, thus allowing it to set up coastal defenses there.

Despite this career setback after the failure in Tsushima, the political winds shifted again in 1862, and new leadership tried to reform the shogunate's military using bannermen like Oguri. In 1862, Oguri began working as a military reformer and finance commissioner, positions in which he set himself apart as a Tokugawa hard-liner. He received military positions as head of the inner guard in 3/1862 and infantry commissioner in 12/1862, a position he held concurrently with that of finance commissioner. As part of the 1862 Bunkyū reforms, Oguri helped reform the infantry based on a Western model, even though he disagreed with the political ideas of the reform's architect, Matsudaira Shungaku, who sought compromise between the court and the shogunate and had begun considering the argument to expel the barbarians.[36] In the fall of 1862, Oguri argued, "The politics [of Japan] have been entrusted to the shogunate since the Kamakura shogunate. Yet recently, not only has Kyoto started meddling, but the daimyo have been speaking out as well."[37] At the same time, a report from Kyoto surfaced that Oguri and others should be removed from office or they would be assassinated for their opposition to the cause of expelling barbarians.[38]

More controversy surrounded Oguri in 1863, when he became an outspoken critic of the proposed détente between the court and the shogunate (*kōbu gattai*). It was in this year that the shogun, Tokugawa Iemochi, made a trip to Kyoto, the first such journey by a shogun since the founding of the Tokugawa shogunate. His visit and marriage to the emperor Kōmei's sister represented a watershed in the imperial court's growing influence. Military planning by Oguri and like-minded bannermen intensified in relation to the court's strength, but reaction against them was equally harsh. On the evening of 3/28/1863, a posted sign in Nihonbashi warned of a plot to assassinate "officials who accept the ideas of Ii (Naosuke) and Andō (Nobuyuki)" and those who "spout the absurdities learned from foreign countries."[39] Oguri was singled out by name. Fear of assassination grew among Oguri's bannerman colleagues in 4/1863, when someone betrayed their plot to send the shogunate's army to Kyoto.[40] Oguri was fired in 8/1863 as part of a larger purge of anti-court, anti-expulsion Tokugawa retainers, putting him at odds with the future shogun, Tokugawa Yoshinobu.

The Challenge of Shogunal Finances

Verny Park, outside of the Yokosuka train station, pays homage to François Léonce Verny, the French engineer hired to build the Yokosuka iron foundry, arsenal, and dry dock.[41] The Verny Commemorative Museum celebrates Yokosuka's origins. A giant three-ton steam hammer imported in 1866, in use until 1996, imparts to the viewer a legacy of industrial modernization from the Tokugawa period to the present. From an initial iron foundry (*seitetsujo*) with a dry dock used to repair Japan's ships and an arsenal, Oguri wanted to see the creation of a shipbuilding facility that would allow Japan to wean itself from dependence on foreign powers for its ships. Yokosuka has been praised for its contributions to victory in naval battles, as a fount of French culture in Japan, as the origin of modern accounting and management, for training a generation of naval engineers, and even as marking the moment when Japan changed from being a country of wood to one of iron.[42] However, others point to Yokosuka as an example of Oguri endangering Japan's sovereignty by depending too much on French financial and technological assistance.

Although Oguri's desire to build a dry dock and iron foundry may have originated from his experiences on the 1860 embassy, the earliest documented evidence of his intentions regarding shipbuilding trace back to the 1861 Tsushima Incident. During a meeting between Oguri and the Russian captain Birilev, the two discussed the price of ships in Asia and the state of cannon and shipbuilding in Japan. Oguri told Birilev, "I think our country should make its own large ships here in Japan. We've already begun doing so in Nagasaki."[43] In 1864, Oguri was assigned to the Nagasaki Iron Foundry during his many on-again, off-again stints as the finance commissioner.[44]

Oguri faced an uphill battle trying to convince the shogunate to commit to the long-term project of constructing a shipbuilding facility. The shogunate and fourteen domains established such facilities during the 1850s after ending the ban on oceangoing ships in 1853, but most were abandoned or lacked adequate facilities to serve as part of a major shipbuilding center.[45] Even Satsuma and Chōshū provinces discovered that their people lacked the technical skills and funding to build their own ships. Oguri also encountered resistance from shogunate officials who feared the project's high cost. Some foreign observers argued that Japan should simply continue to purchase ships from them on the grounds that creating its own shipyard would be too costly and take too long to make it worthwhile.

The shogunate, indeed, could never have embarked on such a project on its own; Oguri and others placed their hopes in the French representatives in Japan. As early as 1864, the French enjoyed a special relationship with shogunate officials when the newly appointed French minister Léon Roches was taken into the shogunate's confidence by then foreign commissioner Takemoto Masao.[46] Oguri hoped to strengthen the shogunate's military in preparation for its attack on Chōshū. After defeating the shogunate's domestic enemies, he then planned to develop Japan's mineral resources by having French technicians expand Japan's mines.[47] The first and most important French-related infrastructure project, however, involved heavy reliance on French engineering and a substantial investment in equipment. Luckily for Oguri, the shogunate had received ship-repairing machinery from the Saga daimyo, who, after purchasing it from the Dutch, realized his domain did not have the financial or technical means to construct a ship-repair facility.[48] Roches enlisted the help of François Verny, an engineer who had successfully constructed a shipyard in Shanghai.

In addition to starting the Yokosuka project, Oguri wrote a contract with the French, securing the money to pay for Yokosuka. The contract provided for the construction of one iron foundry, two ship-repairing stations, three shipbuilding facilities, plus funds for an arsenal and French personnel. At Roches's suggestion, the shogunate also established a school in Yokohama to train Japanese in engineering and French language. Both Oguri and Kurimoto Joun, Oguri's French-speaking assistant in Yokosuka, sent their sons to attend classes.[49] The total cost projected by Verny totaled $2.4 million to be paid over four years.[50]

Yokosuka did not help the shogunate against its enemies, but not due to a lack of productivity. The shipyard completed eight ships by 1868, with eleven more in production, and had the shogunate decided to use the navy to resist the Satchō-led armies, the outcome could have been very different. In any event, Yokosuka was not completed until after Oguri's execution and the shogunate's fall. The Meiji government continued to employ French technicians, despite initial suspicions that the

French might resist due to their support for the shogunate—worries that proved unfounded at Yokosuka.

Oguri pursued two basic strategies to pay for expensive projects like Yokosuka in light of the shogunate's lack of funds. Forcing wealthy merchants to loan the shogunate money had been a time-tested way of obtaining necessary cash, but this had its limits. Oguri drew on his personal connections with the Mitsui merchant group to foster what became a tense but mutually advantageous business relationship. The second strategy involved convincing wealthy commoners to put up the capital for a particular project, with the promise of a share in the profits. The first attempt at this new form of business venture succeeded. In 10/1867, Oguri asked an Edo architect, Shimizu Kisuke, to build a Western-style hotel for the new foreign settlement. According to the diary of Hirano Yajurō, an Edo resident, Oguri planned the hotel to be funded by craftsmen and Edo townsmen, citing the shogunate's own lack of funds to do so. The land would be lent free of charge, and those who invested would receive a share of the profits.[51] The project was a success, attracting one hundred investors. Shimizu completed construction of the American-designed, two-story Tsukiji Hotel in 1868, but it was destroyed by fire in 1872.

Oguri depended on an old acquaintance for his projects: the former family servant and confectioner Minogawa Rihachi. In the late 1850s, when Oguri debased coins in an effort to stop the flow of gold coins leaving the country, Minogawa bought as many old Tenpō coins as his capital would allow at the Mitsui exchange. It is believed that inside information from Oguri helped Minogawa earn windfall profits during the revaluation of official coins, catching the attention of merchants at the Mitsui exchange.[52] Although the Mitsui were apparently loath to enlist the help of an outsider, they hoped to take advantage of the relationship between Minogawa and Oguri to ease their dire financial situation. Minogawa quickly proved himself useful to the group by advising the Mitsui to split their various operations into separate businesses so that failure in one would not affect the others. They appointed him head of the Mitsui's official business establishment (*goyōdokoro*), and he changed his name to Minomura Rizaemon.[53]

The relationship between Oguri and the Mitsui, mediated through Minomura, was a tense one. Oguri needed the Mitsui's money, management expertise, and reputation, while the Mitsui wanted to maintain favorable relations with the shogunate to survive the economic chaos of the mid-nineteenth century. The shogunate coerced loans from many large merchant households, but the Mitsui's portion was among the most burdensome. From 1863 to 1866, they paid approximately 3.5 million *ryō* to the shogunate to fund a variety of expenses.[54] For example, when Oguri launched a second punitive expedition against Chōshū in 1866, he asked the Mitsui to pay 500,000 *ryō* and manage all relief loans in Edo.[55] When Minomura negotiated with Oguri to reduce the burden, Oguri agreed and reduced the payment to 18,000 *ryō*, saving the Mitsui from bankruptcy.[56] Before Minomura's employment, Oguri placed the Mitsui in charge of disbursements of Yokohama custom revenues loaned to Edo merchants—Oguri's solution to the financial panic brought on by the initial Chōshū expedition. Later, when revenues went missing, Oguri forgave the Mitsui of wrongdoing, thus saving the Mitsui's reputation.[57]

In 1866, Oguri solicited contributions from other commoners, such as the wealthy entrepreneur Takai Kōzan, from modern-day Nagano Prefecture. Takai was one of the few wealthy commoners who supported the shogunate, but he hoped the shogunate would reform and recognize the supremacy of the court. The Kujō family in Kyoto was one of his patrons, and Takai believed that the shogunate would be included in a new "union of court and military" (*kōbu gattai*) arrangement advocated by many in Kyoto.[58] In 1866, he met Oguri in Edo and offered the shogunate 10,000 *ryō*, a large sum by any measure, to be paid in annual installments (he paid only 3,000 *ryō* before the shogunate fell). Takai expected that his donation would allow him to voice his own opinions about the shogunate, so he sent Oguri a long list of shogunate reforms.[59] Still, Takai respected Oguri, and he sought advice from him regarding plans to build a local shipping company in Nagano, a plan that never materialized.[60]

Oguri also depended on merchants for organizational support, relying on the Mitsui to help establish the Hyōgo Shōsha Trading Company, his second major effort to employ the sort of profit-sharing arrangement that had produced the Tsukiji Hotel. However, by the time Oguri established an outline for the Hyōgo Shōsha in 1867—his last major project before the end of his career—the shogunate was desperate for money. Through Hyōgo Shōsha, Oguri attempted to find a way to correct fundamental flaws in the trade port system.[61] This is reflected in the language of the Hyōgo Shōsha proposal, written by a joint committee of four men, including Oguri: "it was a loss for us when the foreigners opened the ports of Nagasaki and Yokohama. We had no merchant union system for bringing profit to the government."[62] Hyōgo Shōsha comprised twenty merchants, organized hierarchically by the amount of capital contributed by each. Oguri approved all of the appointments, with instructions on the company's operation given by a government representative who reported directly to Oguri. Hachirōemon, a Mitsui merchant, acted as temporary president of the group due to Oguri's connection with Minomura.[63] Unfortunately for Oguri and the shogunate, Minomura advised Hachirōemon not to invest in Hyōgo Shōsha, after which other merchants hesitated, and the shogunate fell before operations began.[64]

Although Oguri received support from Minomura and Takai, he also bore much of the anger from Edo residents who suffered under the deteriorating economic conditions—in particular, the rising inflation that arose from coinage issues. As prices rose, merchants began hoarding rice, further antagonizing the populace. In 8/1866, a poster appeared in Edo threatening Oguri's life: "Among the government officials are many bad people. Oguri Kōzukenosuke and his colleagues think it is fine that prices rise.... Since we have no other way to save our lives, we are ... going to attack the Oguri mansion in Surugadai. Then we will kill his evil henchmen, save all the people from disaster, dispel our rage, and make the name of the Edokko [Edo natives] known throughout Japan."[65] Threats against Oguri's life never subsided; rumors of his assassination even reached the Western press as late as the spring of 1867.[66]

Advocating War

In the last two years of the shogunate, Oguri occupied positions that allowed him to reform the shogunate's military capacity while building the infrastructure at Yokosuka and attempting to create the Hyōgo Shōsha. He also helped plan the 1866 punitive expedition against Chōshū. Although defeated, the shogun, Tokugawa Yoshinobu, made one last push to strengthen the shogunate, especially its military. To this end, he turned to the French for help. This benefited Oguri's career, as he was made commissioner of the navy, a position he held concurrently with the post of finance commissioner. Oguri used his newfound influence to organize an attack on Edo's Satsuma Mansion in 1867, using French military advisors. Both Oguri and Yoshinobu were excited by the prospect of continued support from the French, but the gap between encouragement from Roches and the French government's commitment to the shogunate ultimately disappointed them.[67] After the shogunate's defeat at Toba-Fushimi, Yoshinobu abandoned hope, even as Oguri tried to press on.

Yoshinobu had already capitulated to the imperial forces, but Oguri led a faction of retainers that argued for a military defense of Edo. Katsu Kaishū argued for a peaceful surrender to avoid bloodshed, thus creating an oft-cited rivalry between Katsu and Oguri, which continues to be a cornerstone for Oguri's commemoration. Officials and daimyo met at Edo Castle from 1/12 to 1/14/1868 to discuss the shogunate's fate. Oguri proposed to use the shogunate's army to fight the emperor's forces at the

Hakone Pass as they advanced along the Tōkaidō highway. There they would trap the emperor's army so that the shogunate's navy could bombard it from Suruga Bay. Afterwards, ships would be dispatched to Hyōgo to halt the advance of the imperial troops.

The factions initially agreed to follow Oguri's plan, but the mood had changed by the following day, and Yoshinobu rejected the idea of military conflict. Katsu had won the day.[68] Ōmura Masujirō, one of the founders of the modern Japanese army, later commented, "We would have been defeated if they had acted on Oguri's plan."[69] Not all of the emperor's men agreed with Ōmura's threat assessment. Etō Shinpei is supposed to have said, "Oguri was that kind of an ass; that's why they didn't buy into his plan."[70] Either way, during this final meeting, Oguri's frustrations boiled over. According to Asahina Masahiro, who attended the meeting, as Yoshinobu turned to leave the room, Oguri grabbed his *hakama* and asked him, "what will your cowardice accomplish?"[71] Without stopping, Yoshinobu relieved Oguri of his official duties, the first time a Tokugawa shogun had fired anyone directly. On the fifteenth, as he writes in his diary, Oguri was called back to the castle, where he was dismissed from his position as infantry commissioner (although he retained his finance commissioner status until his death).[72] Totman suggests that the shogunate made an example of Oguri to warn others who argued for war. After Oguri's departure, many began to favor a more peaceful settlement with the emperor's forces.[73]

Like other Tokugawa retainers, once Yoshinobu decided to surrender, Oguri prepared to leave Edo. Here his diary reveals some of his activities in those final weeks in Edo, namely through the visitors he received. For example, while Minomura is only recorded as having visited the Oguri residence three times before Oguri's dismissal, he visited Oguri twice as much after 1/15/1868. Minomura's son later revealed the goings-on of those meetings: Minomura offered Oguri money, telling him to leave Japan and take refuge in the United States. Oguri refused the money, telling him that he intended to move to his fief in Kōzuke Province, but asked Minomura to take care of his wife and children should anything happen to him.[74] Minomura kept his promise, providing financial assistance and giving the Oguri women a place to stay in the decades following Oguri's execution.

Several Aizu retainers also visited Oguri, indicating his symbolic importance among those who resisted the imperial army. The Aizu domain was loyal to the shogunate and fought with shogunate troops at Toba-Fushimi. Matsudaira Katamori, the Aizu daimyo and protector of Kyoto, returned to his domain in the same month that Oguri met with his retainers, some of whom had recently fought at Toba-Fushimi. Unfortunately, the contents of the meeting between Oguri and the Aizu retainers are unknown. Most likely they reported on their experiences at Toba-Fushimi and discussed possible future action against the Satchō forces. One of the retainers, Akizuki Teijirō, later helped Oguri's wife and mother during their escape from Gonda to seek refuge in Aizu.[75]

The only other insight we have concerning Oguri's post-Edo plans comes from an interview with one of his last visitors, Shibusawa Sei'ichirō, cousin to the famous Shibusawa Eiichi. Sei'ichirō acted as commander-in-chief of the Shōgitai, a pro-Tokugawa militia that fought against the imperial forces in Edo.[76] Hayakawa Keison, a local researcher from Gunma Prefecture, once asked Sei'ichirō why Oguri moved to Gonda Village.[77] Sei'ichirō recounted Oguri's comment:

> I can tell you that we had many plans if the former shogun had decided to fight; unfortunately he did not choose this course of action. If the Northeastern domains come together and fight, they will do so without having a leader and thus have no chance of victory. Moreover, once the fighting subsides there would be an argument over authority, and they would begin fighting each other. We would be unable to unify the country. At that moment, foreign powers could take advantage of the situation, which would lead to disaster. If something should happen, I will place myself under the command of the former shogun, send word to others and alleviate the situation. Should everything go well and there is peace, I will live out my days as a stubborn country bumpkin in the mountains. Therefore, I will avoid all of this mess—I have chosen to live in my fief, Gonda Village. I have no other plans.[78]

Of course, the pitfalls of oral history and memory rightfully make us suspicious of this statement, just as it should with memoirs written during the Meiji period by Katsu Kaishū, Fukuchi Gen'ichirō, and others. But I do not discount this anecdote out of hand simply because it appears as part of historical research conducted by local historians during the Meiji period in the Gunma countryside. First, Oguri's "wait and see" approach to the shogun's intentions is in keeping with documented evidence of his departure from Edo and the strategic location of Gonda Village. Second, other bannermen also retired to the countryside and tried starting new lives as commoners, and like them, Oguri's presence brought disorder to the countryside.

Out of the Frying Pan . . .

When memory activists explain why Oguri chose to live in Gonda Village, they often point to his deep connection to the people there. Oguri invited village youths to Edo for military training, which proved useful when bandits attacked him in Gonda Village, and Satō Tōshichi had long served Oguri in various functions in Edo. But there were other advantages to moving to Gonda. Gonda acted as a node connecting Oguri to his other, smaller fiefs, also located in Kōzuke. Strategically, Gonda was located close to the Usui Pass along the Nakasendō, the main highway leading east from Kyoto through the mountains into the Kantō area and Edo. If the shogun decided to resist the advancing Meiji armies, Oguri could organize a defense of the pass. Gonda was also close to the Aizu domain, where Oguri had acquaintances who could offer help. Moreover, with French technical assistance, the shogunate directed gold-mining experiments in several nearby mountains, resources that Oguri might have wanted to secure for the shogunate.[79]

A key document for evaluating Oguri's post-career intentions is his petition to leave Edo. On 1/28/1868, he submitted a request to move to Gonda, which was approved the following day. His petition stated the following:

> The bannermen and the shogun's lands in Chugoku and Kyūshū are in a state of disarray. If there is anything I can do to help you [Yoshinobu], [remember that] I have experience working as the financial commissioner. I feel so frustrated over what might happen. Luckily I have 2,700 *koku* lands in Kantō, and I will return any amount of those lands as soon as you order me to do so. When I arrive in Gonda, my fief village located in Kōzuke province, I intend to assist in any way that I can. I will establish a peasant militia for when the time comes. If there is something you need, I want to help.[80]

Shirayanagi Natsuo points out a confusing discrepancy regarding Oguri's petition: Oguri offers to return his lands to the shogun while simultaneously informing the shogun of his plans after his relocation to Gonda. The first possibility is that Oguri offered his lands as a formality and did not intend to return them. However, because of the shogun's anger toward Oguri, Oguri presented the shogun with a choice: either you take back my lands or I will leave for Gonda and create a militia to help stabilize the area.[81] Since we know that Oguri was already preparing to leave prior to this petition, it is more likely that he wanted to demonstrate his sincerity to the shogun, perhaps offering him a chance to reconsider armed resistance. Retaking Oguri's lands would have added to the bureaucratic frustrations already facing the collapsed regime. Either way, the shogun declined his offer on the following day.[82]

Oguri left Edo at a time when other Tokugawa retainers were flooding out of the city.[83] The shogunate encouraged fief-holding bannermen to leave for their fiefs not only to relieve some of the pressure of having too many disaffected and now unemployed retainers in Edo but also to disperse its military rather than keep it centralized in Edo, an attempt to convince the imperial army that it had no intention of rebelling.[84] The bannermen worried about their families as they left for the countryside, and continued to mistrust the new government even as the Tōkaidō imperial army tried to placate worried Tokugawa retainers by promising them employment.[85]

Of course, villagers did not always welcome their refugee bannermen, but the grave financial situation for bannermen in the latter half of the Tokugawa period at least allowed villagers room to negotiate.[86] William Steele provides an example of how peasants resisted a bannerman's encroachment on village autonomy in 1868. The villagers demanded that their bannerman reform his financial situation and complained of tax corruption, eventually petitioning not only the shogunate but the new imperial government after the Restoration.[87] Another bannerman, Mishima Masakiyo, ignored his villagers' complaints when he decided to open up a confectionary shop in Edo. They did not want to be responsible for his poor business decisions, and the shop soon failed. Mishima left Edo in the summer of 1868 to avoid persecution by the emperor's armies, and even entered his name as a peasant in the village register. After failing as a farmer, he returned to the capital and failed as an antiques dealer before finally giving up his lands to the government.[88]

By the time Oguri arrived in Gonda in 1868, Kōzuke and the Kantō plain in general had experienced almost a century of disorder. The nineteenth century started badly for the region with the explosion of Mount Asama in 1783, which, combined with the previous year's bad weather, led to the Tenmei famine of 1783–87. Although the region quickly recovered and experienced some degree of prosperity, bad weather destroyed crops, leading to yet another Kantō-wide famine in the 1830s. Lower-class peasants suffered the most, while rural entrepreneurs benefited from the increasingly commercialized economy and the rise of by-employments such as the silk industry, an especially profitable activity in Kōzuke. More peasants were likely to leave their lands and seek opportunities in local castle towns or Edo itself, while poor samurai drifted throughout Kantō and Kōzuke in the mid-nineteenth century. Famine had disrupted peasant lives in Kantō from 1866 to 1868, just as bannermen and daimyo left Edo for their fiefs, while inflation arising from the trade imbalance angered commoners in Kantō and Edo.

The shogunate attempted to regain control of an increasingly unstable population in and around Kōzuke Province in the early and mid-nineteenth century. The Kantō region in general was a mix of bannermen and daimyo fiefs and lands directly managed by the shogunate through intendants.[89] In 1793, the shogunate set up an intendant station in Iwahana

Village near Takasaki—the closest large domain and castle town near Gonda—to strengthen control over the commoner population, especially villagers who had been leaving the area.[90] In 1805, the shogunate established the Kantō Regulatory Patrol, appointing eight men to arrest various troublemakers, especially bandits and unregistered persons in Kōzuke.[91] In 1827, the Kantō Regulatory Patrol expanded its role by organizing villages into groups that would report on any criminals in the area. More importantly, its officers attempted to keep villagers focused on their agricultural activities at a time when more commoners were engaging in handicraft industries.[92] Local representatives of shogunal authority could not manage the amount of violence in the region, so they established a martial arts training center for the sons of peasants, who were supposed to form the core of a newly created peasant militia.[93] The shogunate never completely controlled the Kōzuke countryside, nor did it solve any problems for local officialdom. In fact, villagers bitterly opposed the intendant presence in Kōzuke, particularly because the shogunate forced villages to pay for the station's upkeep. As the finance commissioner, Oguri had been in charge of approving funds for the station, another potential source of local resentment against him.[94]

In the mid-nineteenth century, the Kantō region experienced a type of violence that differed in degree, frequency, and composition from anything seen before in the Tokugawa period. Anti-shogunate, anti-foreign, and pro-imperial terrorism spread from major cities, namely Edo and Yokohama, into local castle towns, such as Takasaki. Shibusawa Eiichi, for example, plotted an attack on Takasaki and the Iwahana station as the first step toward a march on Yokohama, where he would help "expel the barbarians."[95] Economic, political, and social instability gave rise to groups of dispossessed youths who attacked symbols of authority. Unemployed, pro-imperial samurai formed a faction in the Mito domain that targeted anyone sympathetic to the shogunate, including those who did not make donations to their cause. These groups clashed with Takasaki troops in 1864, further destabilizing the region.

Commoner-led uprisings and millenarian protests expanded beyond pre-nineteenth-century patterns of protest and uprisings. The Bushū Outburst, in Saitama Prefecture, spilled into Iwahana and Takasaki from a neighboring province in 1866.[96] So-called "world renewal" riots, attacks on local authority, and "smashings" against local entrepreneurs also increased in the Kantō region. Some scholars see such events as political actions by peasants who "sought to defend local values and local autonomy."[97] Indeed, many scholars of peasant uprisings focus on vertical interaction: poor peasants attacking the rich and other elites. Uprisings before the nineteenth century were planned and well controlled; rarely involved serious violence, such as murder or arson; and forbade fellow participants from carrying weapons. But Suda Tsutomu argues that throughout the nineteenth century, not only did uprisings become more violent but they also did so horizontally.[98] Peasants feared young males whose actions broke with accepted forms of protest. Uprisings in nineteenth-century Kōzuke could at times be interpreted as political and organized, but they could also become violent. Their chaotic nature contributed to an environment of fear that would later plague Gonda-Oguri relations.

The antagonism between pro-shogunate and pro-imperial factions exacerbated conflict in Kōzuke. Shortly before Oguri's departure from Edo, the daimyo and shogunal officials in Kōzuke began preparations to face the advancing imperial army. The Iwahana inspectorate received Yoshinobu's proclamation of 1/5/1868 denouncing the Satsuma domain as traitorous, and distributed it to village officials on 1/13. It ordered village headmen to post the following statement in their village squares or in front of their residences for all to see:

> Of course this applies to Satsuma traitors, but even if there are people who seem suspicious, arrest them with out delay, and if there are not enough people to do this [assist in the arrest], you should kill them [the suspects]. Should anyone hide their intentions to the contrary, they will be punished severely.[99]

The proclamation was an opportunity not just to announce the shogun's attack on Satsuma but also to strengthen control of the countryside.

After the failed battle at Toba-Fushimi, Yoshinobu resigned himself to military defeat and tried to remove his stigma as "enemy of the court," but those in favor of resisting the Satchō troops continued their military preparations. In order to protect Edo from the imperial army's advance through the mountains, the pro-war element needed to defend Hakone and the Usui Pass. Both the Takasaki and Annaka domains set up cannon at Usui, hoping to defeat the oncoming imperial forces. Two armies advanced toward Edo from the southwest: the Tōsandō army, which moved through the mountains, and the Tōkaidō army, which advanced along the east coast. In addition to using loyal domain troops, the Kantō general inspectorate stationed at Iwahana tried to organize thirty peasant militia groups numbering almost fifteen hundred, and arm them with rifles. But some daimyo and bannermen, and most villagers, resisted, so the plan never came to fruition.[100]

The Oguri disturbance in the third month of 1868 and Oguri's subsequent execution on i4/6/1868 illustrate the disorder and ambiguity that constituted the more violent side of the Meiji Restoration. One of Oguri's staunchest supporters was accused of betrayal, his persecutors expressed remorse, and the attitudes of local domains remained usefully unclear. The uprising itself did not fit into any neat category of peasant resistance or political activity, and many rank and file peasants themselves became victims.

The attack on Oguri was a well-planned but poorly executed operation, led by bandits who had been terrorizing much of Kōzuke in the weeks before Oguri moved to Gonda. Within twenty-four hours of Oguri's arrival at the Tōzenji Temple, his temporary residence in Gonda, village officials visited him to express their distress over a group of bandits gathering in nearby Sannokura Village. The bandits threatened to burn the houses of anyone who did not join their cause.[101] Weeks earlier, on 2/19, two bandits by the fierce names of Demon Gold (Onigane) and Demon Queller (Onisada) sent a document to nearby villages during an earlier rampage at the Shimonita post station, telling villagers to send men and bring weapons, including rifles.[102] Onisada joined forces with a Chōshū rōnin named Kanei Sōsuke to form the so-called Oguri Punitive Expedition. Many believed that because Oguri was the last finance commissioner and a military reformer, he would bring with him large sums of money hidden in weapon chests and barrels. The bandits wanted Oguri's putative riches and promised villagers that any money stolen from Oguri would be divided among the participants.[103] Similar uprisings occurred elsewhere in Kōzuke at the end of 2/1868, and wealthy peasants were being extorted for money to lend to the poor.[104] Other groups attempted to gather money for the imperial army's cause. The involvement of pro-emperor rōnin provided a network of information regarding shogunate-related activities in Edo.

Newly relocated to Gonda, Oguri's mind was not far from the politics of Edo, and the controversies surrounding his former occupation followed him. His first day, 3/3/1868, should have been a day of celebration. It was the Doll's Festival, when dolls were floated down rivers, taking with them evil spirits. But Oguri chose not to celebrate: "Today is the seasonal festival, but since the shogun followed orders [from the Meiji government] to remain confined, so did I. There was no celebration today." Other immediate antagonistic forces probably dampened Oguri's mood more than the fate of his lord, however, as bandits were gathering in a nearby town. Under pressure from village officials, he sent Ōi Isojūrō to the bandits' headquarters in Sannokura to "ask about their injustices." Ōi, a

Gonda-born youth, had received education at Oguri's residence and training in French military tactics. Oral history describes Ōi, armed with a sword given to him by his older brother, bravely walking alone through the crowd of rioters. Most secondary sources also claim that Ōi offered 50 *ryō* to the rioters, telling them that Oguri brought with him to Gonda only the amount of money he would need for daily living, and that he did not possess any shogunate funds.[105] Ōi returned that night in the snow, his efforts in Sannokura unsuccessful. That evening Oguri wrote, "It is late in the night, and already I have been planning various defense strategies."[106]

Early in the morning of 3/4/1868, Oguri prepared himself for what would become a bloody day. He sent his mother, wife, and daughter, as well as the women and children of his retainer Mahiko, into the mountains under the protection of headman Satō Tōshichi and retainer Mukasa Kinnosuke. According to the longest entry in his diary:

Mataichi and I were supported by one hundred men made up of my retainers and soldiers, hunters, and other able-bodied men in the village. I divided them into five groups. Mataichi and I led a group of twenty in front of the Tōzenji. I ordered two other groups to defend the flank towards Kawaura Village and sent two more up the mountain (behind Tōzenji). The enemy brought a force of two thousand and attacked from three directions.[107] They set fires north of the temple and in front of a village shrine (one of the rioters' headquarters). Then rifle shots started coming from the other side of the Karasu River near Kawaura Village, so I ordered the soldiers and French-trained troops to advance on the grove around Tsubakina Shrine. The rioters retreated and we chased after them. We shot three men, and a villager killed another with a spear. Once that area quieted down we headed back towards Kawaura but were met with a barrage of rifle fire. We then moved from Ikeda Chōzaemon's house (the other Gonda headman) up to the mountain, not knowing what was happening there. We faced a large group of rioters, but it seemed like Yūzō [Arakawa] and Tōgorō [Kutsukake] (other Oguri retainers) could defend that approach, so I ordered Mataichi to the hills of Kamijuku. We attacked the rioters and chased them from the mountain to the other side of the Karasu River towards Kawaura. Ikeda Denzaburō and Ōi Isojūrō killed about five people while the soldiers killed ten, which routed the enemy. In the mountain, Yūzō and Tōgorō forced the rioters to retreat. I sent Ōi's group as reinforcements, and we gained control over the village, taking several captives.[108]

While Oguri lived in Edo, he became acquainted with a group of Gonda Village youths. He invited them to his residence, where they received education and military training in Edo and Yokohama. Although memory activists tend to romanticize the relationship between Oguri

and these youths, the successful organization and military training illustrated in the previous passage suggests that this was no simple apocryphal village history.[109]

Local people found themselves in an awkward position dealing with
the victorious and suspicious Oguri. Some had clearly fought on Oguri's
side, such as Ōi and Kutsukake, who filled leadership positions in the
battle. In his diary entry on the battle, Oguri singled out Satō Ginjūrō as
being especially praiseworthy: "During today's fight the soldier Ginjūrō
fought extremely well, killing four men. His skills with a rifle are admi-
rable."[110] Other locals had joined the rioters, although their speedy ca-
pitulation indicates that most were passive onlookers at best. Some villagers acted in self-defense, such as one man who speared a rioter, and
others were victims, their houses burned during the battle.[111] Approximately twenty people died in the fighting, some of them villagers caught
in the middle. In one such instance, rioters took over village headman
Maruyama Genbei's house and used it as a base of operations. When
Maruyama returned home from his hiding place after the fighting, he
saw the shape of Oguri's retainer Ikeda Denzaburō and, without thinking, reached for his spear. At that moment, Ikeda beheaded Maruyama,
took his head to show Oguri, and burned down the house.[112] The incident
is recorded in the 1975 village history as "told by Maruyama Daijūrō"—
presumably a descendant or other relative of the decapitated Maruyama—
demonstrating that even well into the twentieth century, the trauma of
this battle still remained in the village's historical memory.

Regardless of Oguri's military strength, his future in the village depended on cooperation with local officialdom. The situation was particularly tense, as several local elites who had fought against Oguri had been
killed in battle, including the village headmen at Miyahara and two village officials from Iwakōri.[113] Nonetheless, on the evening following the
battle, village officials from Kawaura, Iwakōri, Mizunuma, and Sannokura visited Oguri at Tōzenji to apologize: "The bandits manipulated
us and we could do nothing [against them]. We apologize. Let the six-
village cooperative have good relations with you, and if there should be
any differences between us in the future let us talk them through."[114]
Oguri forced one official from each village to stay the night, keeping
them hostage in case any further attacks occurred, but released them on
the following day.[115] They returned to his residence to submit a letter of
apology, and he freed other peasants captured during the fighting. Oguri
also appropriated a technology of power commonly seen in Edo—the
display of severed heads. He placed the heads of those killed in battle on
a rock wall in front of Tōzenji, even that of the unfortunate Maruyama.

These "bodies-as-signs," as Daniel Botsman calls them, informed local people of Oguri's plan to assert his authority in the area, even as the new government tried to do the same.[116]

News of the riot traveled quickly; the log entry for 3/4/1868 at the Annaka post station recorded the previous day's battle "between Oguri and peasant gangs."[117] Although fighting in Gonda had ended, Kōzuke remained unstable; anxiety filled Oguri's life as he tried to settle in the area. Despite having defeated the bandits, he could not capture or kill their leaders. They continued their rampage elsewhere in Kōzuke, and their return remained a constant threat. Moreover, the imperial forces continued their advance into Kōzuke.

Oguri's presence caused local tension, threatening the village's relative self-autonomy and bringing violence to the area, but he was also a source of military know-how. When the rioters moved north into Nakanojō, a town managed by shogunal intendants, officials there asked Oguri for help. On 3/8/1868, two officials visited Oguri, and on 3/9 he dispatched a small group of military advisors to Nakanojō—all Gonda-born soldiers trained in Edo. Officials there gathered money, weapons, and volunteers to form a militia, but they had no fighting experience.[118] They called on Oguri for tactical advice and, according to town financial records, paid Oguri to visit for one evening and advise them on defense.[119] The rioters retreated from Nakanojō, and Oguri's men returned on 3/11.[120] A week later, the village headman from Shimosaida—one of Oguri's fief villages—visited Tōzenji to discuss an unspecified development in the village. Oguri sent Ōi Isojūrō and two other soldiers to Shimosaida to investigate. They returned on 3/23 and reported that the area was inundated with rioters.[121] Oguri may have had adequate defenses to protect himself and been able to offer military services to neighboring villages that faced rioters, but his new role also challenged the imperial army's attempts to establish itself as an authority in Kōzuke.

Oguri also maintained contact with Edo, where conditions for his former acquaintances continued to deteriorate. On the day after the riot in Gonda, he sent an express messenger to his residence in Edo, the Komai house (Oguri's son-in-law's family), and the Kusaka house (Oguri's uncle and the family of Oguri's adopted daughter, Yokiko), informing them of the incident. Eight days after the riot, Oguri recorded, "Today the messenger returned from Edo with a response from the Komai and Kusaka families. Edo has been in chaos due to the advancing imperial army, and support for the shogun has made conditions extreme."[122] Oguri's anxiety intensified as letters and visitors from Edo and fief villages continued to arrive at Tōzenji. The most painful news, however, was the death of Oguri's close friend and distant relative Ōbira Bitchū-no-kami, from whom he had received money before leaving Edo. Members of the Shōgitai had attacked several houses in the area, and Ōbira killed one of them with a rifle before being cut down himself.[123] Oguri had supported the pro-Tokugawa Shōgitai early in its formation, but the disorder caused by the fighting in Edo blurred neat distinctions between friend and foe.

The Tōsandō army, under the command of Iwakura Tomosada, son of the famous Meiji statesman Iwakura Tomomi, began threatening the Kōzuke region late in the first month of 1868. On 2/24/1868, the leaders of the vanguard high command, Nanbu Seitarō and Hara Yasutarō, ordered Takasaki domain officials to prepare lodgings for the imperial army on its way to Edo "to defeat the enemy of the court, Yoshinobu."[124] Unwilling and unable to resist, officials there made the appropriate preparations, and the imperial army arrived on 3/8/1868.

The first task for the imperial army in Kōzuke was to establish itself as the legitimate military and administrative power in the region. It settled in nearby Takasaki only days after Oguri had battled against the rioters in Gonda and chased them out of Nakanojō. Its first military operation involved quelling the peasant uprising in Fujioka Town, where "some peasants ha[d] been attacking innocent people of wealth and stealing their money."[125] Knowing that the Tokugawa cause was lost, the Takasaki daimyo showed his support for the imperial army by donating money, weapons, and ammunition. On 4/12, he made a final symbolic break with the shogunate by changing his name from Matsudaira, received from the Tokugawa, to his family's pre-Tokugawa name, Ōkouchi.

In the middle of 3/1868, the government had ordered all local daimyo to take over guarding bannermen fiefs in order to stabilize Kōzuke and prevent attacks against the rural wealthy.[126] By protecting the wealthy, the financially desperate government hoped to protect future resources. On 4/11, Takasaki officials began managing the newly assigned village protectorates, including two of Oguri's villages, Shimosaida and Yorokubu.

A little over a month after the riot, the imperial army finally moved against Oguri, ordering troops from the Takasaki, Annaka, and Yoshii domains to arrest him. The 4/22 order to Takasaki read:

> Recently Oguri built a fort on his fief (Gonda Village) and has set up a gun battery. He is planning something evil, and reports from various sources are difficult to dismiss. We are certain that he is planning a rebellion. He has wicked intentions towards the emperor and wants to return Yoshinobu to power. We are calling on the domains to arrest him. For the sake of the nation

(*kokka*) we should not forget our unified cooperation and loyalty. If this is impossible please contact headquarters at once. We will send several units to destroy him in one swift motion.[127]

The orders alluded to two pieces of evidence that Oguri was planning a resistance against imperial troops: (1) Oguri's possession of a cannon and (2) his construction of a military installation. After the riot, rumors spread that Oguri used a cannon during the fight, but there is no evidence to support this. Oguri owned at least one cannon, used as decoration in his Edo compound, but according to his diary, it did not arrive until a month later, on 4/8/1868.[128] The second accusation concerned the construction of his new residence on the top of Mount Kannon. The small mountain overlooked the major road leading from Takasaki City, giving him a strategic vantage point of the area around his lands. From the end of the riot to the arrival of imperial troops, Oguri spent almost every morning or afternoon visiting Mount Kannon to check on the construction of his house.

With their lord absent, the senior officials in Takasaki frantically consulted with one another and decided to obey the order. Oguri heard about the expedition seven days later and dispatched his Gonda retainers Kutsukake and Ōi to Takasaki, probably to plead on his behalf. They returned prematurely, however, after troops stopped them in Haruna, a town on the way to Takasaki, and abruptly informed them, "We will arrive in Gonda tomorrow to discuss things further."[129] A combined force of nearly eight hundred soldiers set up camp at Sannokura Village, just as the rioters had done months earlier.

Takasaki soldiers interrogated Oguri at his temporary residence. First, they showed him their order, which he later transcribed into his diary. Oguri inquired, "If your orders are true, then no matter how many explanations I give these suspicions will not be dispelled. But you have investigated the area; what do you think is the truth?" The men from Takasaki conceded that there was no truth to the accusations. As for Mount Kannon, Oguri explained that he was having a house built and arable lands cleared for cultivation, nothing that could be considered a stronghold. He invited them to investigate carefully to erase any doubts they might have and then report back to their respective domains. They responded, "We will do so. Our domains really never had any doubts about this (Oguri's plotting), but we would like to take back some kind of proof." Oguri even offered the soldiers his cannon.[130]

The domain soldiers carried out their orders, and the issue appeared resolved. Oguri's wife, mother, and daughter, who had taken refuge at the Ikeda house north of Gonda, returned later that evening. On the following day, i4/2/1868, the date of the last entry in Oguri's diary, the soldiers returned to Takasaki with Mataichi and several of Oguri's retainers, who wanted to reassure the Tōsandō high command of their lord's innocence.[131] Furious that the soldiers did not arrest Oguri, the Tōsandō army command ordered them to return to Gonda and complete their arrest. This time, they were led by Hara Yasutarō and Toyonaga Kan'ichirō, vice-circulating inspectors of Kōzuke for the imperial army.[132]

What occurred between Oguri's last diary entry on the second and his execution on the sixth is unclear, but it appears that he suspected further trouble. According to local oral history, Oguri was told that troops were returning to Gonda, and he planned to have his family temporarily seek refuge in a hamlet called Kamezawa, northeast of Gonda. If Oguri was out of danger, then his family would return the next day; if not, he ordered his retainers to take his family to Aizu. In one version of the story, Oguri went with them to Kamezawa and returned at the request of Satō Tōshichi, who convinced Oguri that meeting with the local authority was in the best interests of the villagers and Mataichi.[133] Another version claims that Satō stopped Oguri before reaching Kamezawa by convincing him that the soldiers had left the area.[134] In either scenario, Satō seems to have betrayed his lord. Moreover, Satō had borrowed large sums of money from Oguri on three occasions: 200 *ryō* on 3/2/1868, which Satō agreed to pay back monthly with 10 percent interest; 100 *ryō* for various farm equipment on 4/2/1868; and another 200 *ryō* on 4/25. It would have been difficult if not impossible for Satō to repay Oguri, providing good reason for suspicious minds to assume that either Satō knew of Oguri's imminent arrest or perhaps assisted the imperial army.[135] According to local legend, as Oguri waited on the execution ground, he looked into Satō's eyes and said, "even after I die I will curse you."[136]

Despite the lack of evidence against Oguri, imperial loyalists beheaded him and two retainers in front of peasants from the surrounding villages. Oguri's adopted son and several other retainers were executed the next day in Takasaki, none of them knowing that their lord had been beheaded the previous day. Oguri's surviving retainers escorted his pregnant wife and mother, and the wives and daughters of several other retainers, to the Aizu domain. There, several of Oguri's men died alongside Aizu samurai fighting imperial troops. Amid the violence, Oguri's wife gave birth to a daughter named Kuniko.

The executions of Oguri and his retainers were not part of imperial government policy. The Tōsandō army's leadership was weak, led by a teenager who depended on the advice of Itagaki Taisuke, a maverick often dismissive of orders sent by Saigō Takamori, the leader of the emperor's forces.[137] And they were quite vicious, executing Sagara Sōzō (leader of the "fake army" fighting in the emperor's name) on 3/3 and Kondō Isami (captain of the pro-shogunate police group called the "Shinsengumi") on 4/25. Panic might have also played a role in Oguri's haphazard execution. In the early twentieth century, the monk from the Fumon'in interviewed Hara Yasutarō, one of the men in charge of Oguri's arrest. Hara stated that they feared Oguri and needed to deal with him quickly because riots had broken out in other parts of Kōzuke, a reference to uprisings in nearby Utsunomiya and Mikuni Pass.[138] Hara, twenty-two at the time, and his sixteen-year-old co-leader, Toyonaga, were simply following Tomosada's orders. Apparently Tomosada also feared Oguri, especially his possession of cannon. Hara regretted taking part in Oguri's death and explained, "In those days people thought winning meant doing whatever one could to kill the other guy first. There was no idea of imperial army versus rebel army; that notion was a Meiji period invention. Each side believed it was the emperor's army."[139]

The Oguri Family Escape

The Oguri family was most likely not under any real threat after Oguri's execution, but they certainly felt they were potential targets. Oguri had asked his executioners to spare his family, and Hara and Toyonaga assured him no harm would come to them. After the execution, the Tōsandō army posted a sign pardoning Oguri's wife, his children, and his remaining acquaintances.[140] Nevertheless, the Oguri women, retainers, and servants retreated to the pro-Tokugawa Aizu domain.

Before his arrest and execution, Oguri met with village officials to plan his family's escape. Returning to Edo might have put the Oguri family and others in danger, as most of his former colleagues had left the city to flee the growing unrest there. Aizu seemed the only natural choice—Oguri had acquaintances there, who formed a sizable army capable of defending his family. When the combined force of Annaka, Yoshii, and Takasaki domain soldiers returned to Gonda under the command of Hara Yasutarō, Oguri's family had already left Tōzenji; they were hiding at the Ikeda residence in nearby Kaminokubo awaiting word of Oguri's fate. Upon learning of their patriarch's death, the Oguri family traveled to Aizu escorted by twenty-one villagers led by Nakajima Sanzaemon, a member of the Gonda elite. Nakajima's seventeen-year-old daughter was assigned to care for Oguri's wife and act as a nurse when the baby was born.[141] Most of the other villagers included those who had fought alongside Oguri during the riot.

Although all of the escorts and the Oguri women arrived in Aizu safely, several incidents help us understand the difficulty of running from the imperial forces. Oguri's elderly mother and wife, Michiko, who was eight months pregnant, had to be carried in a palanquin at different stages of the trip. At one point a Yoshii soldier accidentally discovered the group. Instead of turning them in, however, he hid Oguri's wife in a large grass basket and gave her money.[142] The group lacked cash, and at one point Nakajima sent runners back to Gonda to acquire funds from Satō Tōshichi. They returned empty-handed, Satō claiming he had given all his borrowed money to the army.[143]

Although the Oguri family arrived safely in Aizu before eventually settling in Tokyo, the tragic story of Tsukamoto Mahiko's family, as retold in present-day Oguri lore, serves to highlight the sense of trauma created by fear of the new government. Tsukamoto was originally a retainer of the Tatebe family, of which Oguri's wife was a member. When Michiko married into the Oguri family, Tsukamoto moved with her and served under Oguri. He was beheaded along with several other retainers in Takasaki. The remaining six-member Tsukamoto family (mother, wife, three daughters, and baby boy) only accompanied the Oguri family for a short period before turning south for the Nanokaichi domain, where they hoped to receive aid from a relative who worked in the Edo castle among the women of the "great interior."[144] Maki, Tsukamoto's mother, apparently decided that the family should split up again, perhaps worried that the group's size inhibited their movement through the mountains. Maki took her eldest granddaughter, Chika, while Tsukamoto's wife, Mitsu, cared for her two daughters and baby boy. While traveling along the Jizō mountain road toward Nanokaichi, Maki encountered a local man, Shimoda Kijurō, who promised to guide her but then left to bring them food. Maki, possibly worried that he had alerted the authorities, slit Chika's throat and then killed herself.[145] Meanwhile, Mitsu and her three children became lost in the mountains. She then appeared with her son at the Takei household and asked for help. According to the Takei family, Mitsu had drowned her two daughters, fearing that they would not last through the journey. She then gave the Takei a short sword for helping her, which has become part of the material memory of Oguri spread across the Kantō countryside.[146]

Aizu, the Oguri family's final destination, suffered the designation of "enemy of the court," and became a major site of the Boshin War. The group arrived at the residence of the Aizu domain elder, Yokoyama Chikara, at the end of the fourth intercalary month. Although Oguri's diary does not mention Yokoyama, the two probably knew each other through their mutual connection to the French; Yokoyama was an exchange student in Paris during the 1867 World Fair. When the group arrived at Shirakawa Castle, Yokoyama was already fighting Satchō armies at Shirakawaguchi. Their host did not live long, and in the beginning of the fifth month, Yokoyama's head was delivered to his family at Shirakawa Castle after Aizu troops failed to protect the pass to the castle. The Oguri and surviving Yokoyama family members escaped to Wakamatsu Castle. Not long after, Oguri's mother and wife fled the ensuing devastation at Wakamatsu, Michiko giving birth to a baby girl in a makeshift battlefield hospital.[147] As for Oguri's relatives, his cousin Kusaka Junosuke acknowledged Oguri's execution and, after declaring his loyalty to the emperor, requested that he follow the other Tokugawa retainers to the Tokugawa homeland in Shizuoka Prefecture.[148]

Coda

Several years ago, I contacted an American scholar of Japanese history who, during the 1970s, considered writing a dissertation on Oguri. His advisor at the time discouraged him from doing so because, the advisor reasoned, there were too few sources to do the topic justice. Oguri left behind only a few documents; moreover, as a commissioner, many official documents bearing his name are also signed by others, making it difficult to gauge which policies were completely his own or originated elsewhere. Later commentators used this ambiguity to either praise or criticize Oguri and, by extension, the shogunate. For example, although Oguri is typically credited for the Yokosuka project, some historians wonder if it was really Roches's idea, a French attempt to force its interests on Japan. The most well-documented events in Oguri's life were the most controversial, and those fault lines in his biography have formed the basis for conflicting interpretations down to the present.

In the decades following the Boshin War, many of these interpretations came from former colleagues who recalled their interactions with Oguri. Those memories tell us something about Oguri's personality. However, they need to be used with caution, not because they are false or fabricated but because they are recalled only after self-editing, revealing more about the commentator than giving a complete and accurate depiction of Oguri. Take, for instance, his work on Yokosuka. Among all of Kurimoto Joun's dealings with Oguri, Kurimoto chose this often-cited anecdote to summarize Oguri's attitude about Yokosuka's high cost and the shogunate's future:

> I said to him [Oguri,] "the expense is enormous, let's really take everything into account. . . . Once we sign a contract [with the French] then we can do nothing." He laughed and replied, "The economy of our time is essentially a makeshift one. For example, if we don't use money for this [Yokosuka], and instead use it for something else, well, the fact is there is nothing else. There-fore, it would be better to construct this much-needed shipbuilding dock, rather than saving the money only to waste it later. And, once it is completed, even if someone else takes it over after, we would have the honor of selling a house [the government] with a storehouse attached.[149]

For Kurimoto, this quote demonstrates Oguri's foresight that the sho-gunate would fall, and his desire to see the project through even if the regime he served fell—the depiction of a true patriot.

Some stages of Oguri's life and career are known solely through anec-dotes. The only story about his childhood, probably apocryphal, comes from a biography written in 1928 by a descendant of Oguri's wife's natal family, the Tatebe. His schoolmates supposedly nicknamed him "Tengu" because he did not suffer fools and fearlessly offered up his opinion. In one unlikely scenario, a fourteen-year-old Oguri debated his future father-in-law, Tatebe Masatsu, a neighbor in Surugadai: "He was like an adult; his words were clear, his voice booming, and his attitude striking as if already a samurai retainer. Although only fourteen, he smoked tobacco and struck the ashtray with his pipe while answering Tatebe's questions one by one. While everyone was shocked by his haughtiness, they won-dered what kind of person he would become in the future."[150] Resolute, opinionated, and smoking all the while, Oguri appeared as the very image of a 1920s "modern boy." The reader of this passage in the biography has a sense that Oguri was an intellectual, ahead of his time, but also possessed a confidence that ignored status sensibilities.

Memoirs about Oguri do not simply remind people how great he was but also convey a sense of the anger felt by those who disliked him, including many pro-emperor, anti-Western zealots, for whom he was a target; discontented commoners; and even some colleagues within the shogunate. For example, in order to cut wasteful spending, Oguri appar-ently wanted to eliminate monetary gifts given to shogunate officials. Togawa Zanka tells us that Oguri cut shogunal expenses by curtailing bannermen salaries, reducing the budget of the castle women (*ōoku*), and firing priest servants (*bōzu*), who responded by punching holes in the *shōji* screen doors throughout the castle.[151]

Like most bannermen, even those who served as commissioners, Oguri might have slipped into historical obscurity. But by a combination of historical accident and his own policy efforts during key moments in the shogunate's history, he was later deemed important by those who commented on and studied the closing years of the shogunate. Whether Oguri's role was considered positive or negative depended on the speaker, his or her agenda, and shifts in historical contexts.

CHAPTER 2

Creating Tokugawa Heroes in Meiji Japan

Tokugawa retainers suffered during the Meiji transition years—some died fighting against the Meiji forces in Edo, the northeastern domains, and during the final battle at the star-shaped fort (*goryōkaku*) in Hakodate. Most chose not to fight, and many of them served the Meiji government, but many others found it difficult to secure new employment. Some tried to open shops, often to sell off their own family possessions, and others took up farming. When the last shogun retired to his clan's homeland in Shizuoka, thousands of retainers and their families followed him. Samurai refugees left behind property in Edo, quickly confiscated by the Meiji government; their loss of privilege presaged the larger identity crisis all samurai would experience during the 1870s. And when the last resistance against the Meiji army was defeated, Tokugawa supporters lost their ability to wage a "war of words" through news reporting. Memory and commemoration became their new battlefront. Some former retainers pursued successful careers in the world of letters, becoming prolific journalists, writers, historians, and educators. They continued to criticize the Meiji government but no longer as Tokugawa loyalists.

The emerging historical narrative promoted by the new Meiji government branded the Tokugawa regime and its supporters as backwards, at best, or enemies of the emperor, at worst. In response, ex-shogunate men created their own narratives that acknowledged the faults of the Tokugawa era yet challenged the notion that the new Meiji government alone was responsible for Japan's development. By demonstrating the Tokugawa regime's role in the formation of modern Japan, they attempted to rescue the Tokugawa legacy, and their own, from new dominant narratives. They used Oguri's legacy as a tool to criticize both the Meiji oligarchy and fellow Tokugawa retainers who worked for the new government. In the process, they turned Oguri into a martyr. As Michel de Certeau noted, the martyrdom tale is predominant wherever a community is marginal or threatened with extinction.[1]

Certeau points to what he calls the two "apparently contrary" movements that are connected through the narrative of the "Life of the Saint." On the one hand, a group distances itself from its origins, defining itself by the shift that marks it as different from the past. Thus, former Tokugawa men accepted that the shogunate was feudal, and they portrayed the move away from the shogunate-domain system as a good one. Yet when a group risks being dispersed, Certeau argues further, a return to origins reestablishes unity.[2] By writing alternative histories and holding Edo commemorative celebrations, Tokugawa retainers ensured that government-sponsored historical discourse would not obliterate their legacy. Oguri proved doubly convenient for this task, not only as a martyr but as a "Mikawa samurai" whose family connection to Tokugawa Ieyasu, the shogunate's originator, made him the ideal hero for uniting Tokugawa men. By rehabilitating Oguri, Tokugawa "losers" sought to redeem their own value to society and clarify their role in history.

This chapter examines the appropriations of Oguri within the larger context of Meiji-era historical memory and the production of historical discourse at the national and local levels. During this time, former colleagues, many of them national figures, dominated writings about Oguri. Even before the state histories accepted at face value the charge against Oguri that he was an enemy of the court, newspaper accounts immediately after his death set the stage for his early appropriation in Tokugawa apologetic works.

Few biographies or research focusing on former shogunal officials were published during the early Meiji period, except about those who became major figures.[3] Oguri's brief appearances in writings at the national level did, however, help establish a base for the textually mediated collective memory about him: the stock of stories, facts, and interpretations that informed subsequent memories of the past, despite the contested nature of those memories.[4] Later memory activists—or "reputational entrepreneurs," as historical sociologists term them—never worked from a blank slate when appropriating Oguri and other historical figures but addressed the "reputational trajectory" established early in a particular figure's commemoration.[5]

The most influential biography was written by Tsukagoshi Yoshitarō, a member of the village elite from the area where Oguri was killed. Tsukagoshi acted as a bridge between national and local historical memory of Oguri and the Meiji Restoration. He joined the Min'yūsha intellectuals in Tokyo and used his newfound position to promote Oguri as a national hero. He adopted much from former Oguri colleagues, such as Fukuchi Gen'ichirō, but as a native of the place of Oguri's death, his writings about Oguri tapped into local knowledge that was largely unavailable to writers in Tokyo.

While pro- and anti-shogunate figures debated Oguri's legacy in print, in Kōzuke his memory was mediated not only through texts but also through what Jan Assmann and John Czaplicka call "communicative memory."[6] In contrast to memory that has permanency through the written word, communicative memory is disorganized, unstable, and nonspecialized. This memory lives in everyday communication through rumors, jokes, and gossip. Thanks to the controversies surrounding Oguri's presence in Kōzuke, few locals chose to valorize him. People in Kōzuke remembered him largely through superstition and legends—in particular, the buried-treasure stories that placed Oguri within the violent transition experienced during the creation of Iwahana Prefecture, the earliest manifestation of Gunma Prefecture.

Structurally, this chapter is divided into two categories of memory that tended to dominate in two different types of locations. Textually mediated memory initially appeared almost entirely at the national level. During the 1890s, however, textually mediated memory did appear in the countryside. Moreover, early pro-Tokugawa newspapers in Edo recounted rumors that characterize communicative memory. But in the Gunma countryside, in areas affected the most by the disorder of the Tokugawa-Meiji transition, communicative memory lasted longer. In the case of Oguri's legacy, communicative memory persisted even after he became a regional hero and memory activists sought to monopolize the narrative about him.

Part One: Remembering the Restoration in National Historiography

The importance of history production was evident from the beginning of the Meiji period, when an 1869 imperial rescript emphasized the need for an official history. Establishing an official history served the government's need to articulate the relationship between the emperor and his people and, in so doing, to vilify previous centuries' rule by warriors: "Now the evil of the misrule by the warriors since the Kamakura period has been overcome and imperial government has been restored. Therefore we wish that an office of historiography be established, and the good custom of our ancestors be resumed."[7] Government historians who worked during the first several decades of the Meiji era tried to continue the historiographical traditions begun in the *Six National Histories* (*Rikkokushi*), which left off in year 886, and the *Great Chronological History of Japan* (*Dai Nihon hennenshi*), which ended in 1392. The nature and location of the government history offices changed several times during this early period, but two consistent features emerged. First, most of the officials came from pro-Meiji domains, and thus many were considered "loyalists" according to early biographies.[8] Second, these officials followed a tradition of data collection and textual criticism that neither attracted readers' interest nor completely satisfied the political goals of the time.[9] Following the German model, they attempted to produce factual knowledge "but [lost] sight of the inner connection between historical thinking and contemporary experience and problems of orientation."[10] Although such state histories became important primary sources for later historians, they failed to move an elite public readership, as did the narrative-based works by former shogunate men such as Fukuzawa or Taguchi Ukichi, whose goals were the political enlightenment of their audience.[11]

The official historians managed, moreover, to anger those both in and out of the government. Many intellectuals outside the government criticized the project for being entirely in Chinese, so that it not only was beyond the reach of the average reader but also flew in the face of nationalist sentiment—why write a domestic history in a foreign language? Nor did nationalists appreciate the historicizing of old heroes who, upon closer scrutiny, were shown never to have existed.[12]

In addition to the ambitious project of writing a comprehensive national history for both domestic and foreign consumption, the government sought to produce an account of the Meiji Restoration and assigned officials the task of gathering materials. Early efforts were met with hesitation by Itō Hirobumi, who feared that writing a history of the Meiji Restoration might exacerbate tensions between the Satsuma and Chōshū oligarchs, whose home domains had once been enemies.[13] Eventually finished in 1889, the official account—*Record of the Restoration* (*Fukkoki*)—was a long collection of documents outlining a chronology of Restoration events. Like other state history projects, it lacked a narrative and took too long to complete to have much influence on a popular audience.[14] Although completed in the same year that the Meiji Constitution was announced, it did not reach a mass audience until 1930, when it was finally published in printed form.[15]

While professional historians who worked for the government concentrated their efforts on document collection and textual criticism, private scholars and journalists dominated narrative history.[16] Many of these producers of historical narratives had worked for the Tokugawa shogunate and were critical of the pro-Meiji government bias present in official histories. During the Meiji period, most pro-Tokugawa discourse was introduced through journals, memoirs, and narrative histories, and helped propagate the collective memory of former Tokugawa officials. Their writings were more readable than state-sponsored history projects that eschewed individual experience or opinion. Susan Crane notes that history's professionalization "made it incumbent upon practicing historians to retract almost all vestiges of personal memories or personal involvement in the production of the history."[17] The many unofficial histories written by those who witnessed the events of the Restoration gained an air of legitimacy because their authors could claim lived experience. Even though resistance against the Meiji government ended in 1869, the battles on the discursive front continued. Former Tokugawa men wanted to protect their legacy, refusing to be damned through marginalization or, as in many pro-Meiji histories, mislabeled as "backwards" or representatives of "the evil past." They also wanted to show their contributions to Japan's civilization and demonstrate their support of the emperor system.[18] Thus, pro-Tokugawa histories, far from criticizing the emperor or the imperial institution, even though they were the central features of state-sponsored histories, instead exhibited what Totman calls the "Meiji bias," a strategy of rehabilitating shogunal officials by demonstrating how they were, in fact, patriotic and loyal to the Meiji emperor.[19] For example, *The Seven Year History* (*Shichinenshi*) portrays the Aizu lord Matsudaira Katamori as a supporter of the emperor, one who advocated the union of court and shogunate (*kōbu gattai*) and was the protector of Kyoto. His downfall is depicted as being due to both his allegiance to the Tokugawa and the influence of bad shogunate retainers.[20] These Meiji critics shared with their foes a sense of national unity and the need to articulate Japanese identity, but they also wanted to preempt versions of history that would marginalize the shogunal losers.[21]

Oguri as Martyr

Even before the first pro-Tokugawa histories appeared during the 1880s, an earlier "war of words" had occurred in newspaper reporting. Huffman's thorough study of Meiji Restoration journalism demonstrates that early newspapers were overwhelmingly political and largely pro-shogunate.[22]

Roughly twenty in number, they narrated the Restoration for their readers and criticized the imperial government through essays and cartoons. Typically, these newspapers included reports from the battlefront, announcements from the Meiji government, and rumors about fighting in and around Edo. Reports of those killed were common, but only in vague numbers; people were rarely named. Oguri and Kondō Isami, captain of the Shinsengumi, were the two notable exceptions. In Kondō's case, his execution and the displaying of his head were recorded, along with the signpost that accused him of crimes against the emperor's army.[23] The accounts about Oguri included the uprisings against him; the investigations by the Annaka, Takasaki, and Yoshii domains; his execution; a list of his crimes; and responses of the Oguri family representative and others. One example of an Oguri-related rumor was listed as "a story heard from a man who passed by the Iwahana Intendant Station." It included not only the attacks on Oguri but a description of the fortress he was building behind the Tōzenji Temple. Interestingly, the man also noted the tens of thousands of *ryō* confiscated by the authorities in Annaka and Takasaki—rumored wealth that might have contributed to later buried-treasure legends.[24]

The dominant and longest-running *bakumatsu* newspaper, Yanagawa Shunsan's *Chūgai shinbun*, was the first to mention Oguri. The passage is short:

> Recently Itakura Iga no kami [Katsukiyo] has gone missing. What has happened to Oguri Kōzukenosuke, and the natives on his fief lands who rose against him? Kondō Isami has taken flight and his whereabouts are unknown. Moreover, many famous swordsmen, scholars of western studies, doctors and the like are leaving [Edo] and living elsewhere.[25]

To the average reader, most of them Edo residents, these men needed no introduction. Itakura served as the elder councilor who fought against the Satchō forces in Hakodate in 1868. He was imprisoned but later released and served as the Shinto priest at the Tōshōgū Shrine in Tokyo. Kondō Isami had been executed by the Tōsandō army shortly before Oguri's death. The author's worried tone bemoans the exodus of elite citizens and wonders what will happen to the shogunate's more respected officials, with whom his readers presumably sympathized.

Several common features appear throughout the various reports about Oguri, each affecting subsequent appropriations of him. First, most newspapers only mentioned Oguri after his execution in Kōzuke and presented the same narrative of his death. Second, *Chūgai shinbun*, *Soyofuku kaze*, and the pro-Meiji Osaka newspaper *Naigai shinpō* all used the same document submitted by Oguri Niemon to the government acknowledging the execution of his relative, Oguri Tadamasa. Niemon, from a separate branch of the Oguri family, temporarily represented the entire Oguri family after the execution of the main branch's father and son.[26] Despite slight differences in punctuation and spelling of names, the document in each newspaper reads as follows:

> Oguri Kōzukenosuke arrived in his fief land of Gonda Village on 2/28 [1868] as he had requested [from the shogunate] on the first month of the year. On the 29th of last month, the imperial army of Matsudaira Ukyōnosuke [lord of Takasaki], Itakura Kazue [lord of Annaka] and Matsudaira Kanemaru [lord of Yoshii], totaling three hundred soldiers, went to Sannokura Village. On the first of this month [third month] the group was sent to Gonda Village at the command of Iwakura [Tomosada] to subjugate Oguri and his son. They told Oguri to hand over his weapons and he complied. His heir Mataichi left for Takasaki to further negotiate the situation. . . . [H]e arrived on the fourth at the seventh hour in the morning [four a.m.] with three retainers and three servants and they stayed in an inn. The governor [Iwakura] ordered that their weapons be confiscated, and on the seventh they were taken to the Takasaki commissioner. Mataichi was executed first, followed by the three retainers—they did not even receive an investigation. The three servants were jailed then released. They encountered some villagers from Gonda and asked about the well-being of Kōzukenosuke [Oguri Tadamasa]. Oguri and three retainers were executed at the fourth hour in the morning [ten a.m.] of the sixth without having been investigated. All of his household possessions were confiscated and no one knows their whereabouts.[27] One of the servants immediately reported this to Edo. Intercalary fourth month, twentieth day, 1868.[28]

There are two sets of protagonists in this story: Oguri, and the retainers led by his son, Mataichi. The reader is informed that Oguri followed proper procedure for leaving Edo and safely arrived in Gonda Village. There is no explanation about why he left Edo or why the imperial army executed him.[29] The second part of the document discusses Mataichi's and Oguri's retainers; both sections share the twice-repeated phrase "without being investigated." By using the phrase twice, the author draws equal attention to Oguri's execution and that of his son and retainers, thereby highlighting the government's double injustices. Many in Edo understood the long-held animosity toward Oguri. Thus, his death, though tragic, might not have been a surprise. But by leading the story with the execution of his young heir and retainers, the type of low-ranking men who intermingled with Edo commoners, the author closes the emotional distance between the story of Oguri's death and a more general audience. He suggests the Meiji forces went too far.

Fukuchi Gen'ichirō, the premier journalist in Edo at the time, devoted the most attention to Oguri's death. Like Oguri, he envisioned a reformed, Tokugawa-led government. Fukuchi grew up in Nagasaki; studied Dutch with Nomura, one of the interpreters sent to the United States in 1860; and joined the Foreign Ministry in 1859. He interacted with Oguri, and they shared common views regarding the future of the shogunate. In 1867, after Yoshinobu resigned as shogun, Fukuchi sent a memorial to Oguri outlining a reform that would change the shogunate to a parliamentary system with the Tokugawa as its leader.[30] Ultimately, Oguri did not send Fukuchi's plan through higher channels within the bureaucracy. Although he sympathized with Fukuchi's views, Oguri did not trust the shogunate's offices in Kyoto, where the shogun had been staying, nor believe that such an idea would be accepted there.[31] Both men wanted the last shogun to pursue a military solution to the Satchō challenge, and each developed separate military strategies for fighting the Satchō forces.[32] Shared occupational experiences and political views explain why Fukuchi later became one of Oguri's strongest rehabilitators.

Of the various Restoration newspapers, Fukuchi's *Kōko shinbun*, written simply enough for a wide audience and offering the broadest categories of news coverage, is regarded as Japan's first modern newspaper.[33] Moreover, his commentary was the most critical of the new government and the Satchō forces. After Oguri Niemon's document appeared in the *Kōko shinbun*, Fukuchi commented in a postscript:

> Oguri was a resolute person who put public affairs before his own. He faced all the problems of the nation and never gave up. However, he was stubborn and did not get along well with others. Still, hearing about his death, I feel that the imperial nation lost someone important. Moreover, they did not investigate his crime, they just slaughtered him and I do not know why. They did not value his abilities, nor did they express remorse for this loyal subject. In particular, they did not even consider the people's opinions. I ask you the public of Japan about this [whether it is good]? This opinion piece was sent to the newspaper office under a pseudonym and I [Fukuchi] have printed it here.[34]

Several days later, Fukuchi published an untitled poem that putatively conveyed Oguri's last words. There is no introduction to the poem, and the author is "unknown": "Why have I been accused, they kill us [the family] like pigs, why has heaven allowed this, even though I will go to *yomi* my spirit will never die."[35]

These were the first commentaries published about Oguri's execution and represent the first appropriation of his death. The indignation they express was heightened by the double insult of Oguri's rough handling by his executioners and the public exposure surrounding his punishment. In Edo period punishments, the shogunate did not typically question high-ranking samurai, for to do so equated them with common criminals.[36] Samurai criminals were executed in private using "ritual suicide," receiving more careful treatment than commoners.

In addition to the evident anger expressed over the execution, Fukuchi's commentary and poem imply that the Meiji oligarchs would come to regret Oguri's death because they had failed to value his abilities. This was more than simple hyperbole; the newly formed government desperately searched for talented men who possessed intimate knowledge of the West, summoning many former Tokugawa officials to work in its new ministries.[37] Fukuchi uses rhetoric to imply that the execution ignored the "people's feelings" and thus counter the government's claims that it cared for the good of the people.

Other early journalists cited Oguri's death in criticizing the government for the disorder it had unleashed. In the days after the collapse of the Tokugawa regime, the Meiji government had not yet established any control over the Kantō countryside nor been able to rein in the Tōsandō army. Hashizume Kan'ichi, the editor of the pro-Tokugawa newspaper *Kōshi zappō*, expressed his opinion of these failures in a short article entitled, "For Some Reason the Imperial Army Killed Oguri in Haste." Hashizume accuses Oguri's persecutors of not following the emperor's orders despite claiming otherwise, adding, "not one person should be executed haphazardly if their guilt is not clear. . . . [T]his is truly regrettable."[38]

Ultimately, after the Restoration, the Council of State eliminated most pro-Tokugawa newspapers by issuing a decree requiring all newspapers to obtain official recognition.[39] Fukuchi was jailed for his various attacks on the government but was soon released to continue his career in journalism and become the best known of the Tokugawa apologists.

Rehabilitating Oguri and the Defeated

In general, most narrative histories by former Tokugawa retainers, though part memoir, were also part political commentary, and thus Oguri appeared in them as either a tool for critique or an exemplar of tenacity, hard work, self-sacrifice, and patriotic spirit—qualities highlighted in his biography precisely because he had been purified by death and was above the base politics of the day. It was in this sphere where the first fully developed writings about Oguri appeared. Nationally renowned commentators who wrote about Oguri the most, men such as Fukuchi Gen'ichirō, Fukuzawa Yukichi, Katsu Kaishū, and Kurimoto Joun, had known him well, in particular through their shared involvement with Westerners.[40] Yet as former retainers, they were also connected to one another, sometimes as rivals, a nontrivial issue when considering why some were motivated to write about Oguri in the way they did.

Each author had his own critique of the clique government, being either against the entire oligarchy itself, in favor of "people's rights," or against certain factions within the government. The first author to appropriate a "loser" of the Restoration years for such use in a political critique was Shimada Saburō. Shimada was born in 1852 to a Tokugawa retainer and later attended Numazu Military Academy in Shizuoka, a school built and dominated by ex-retainers in the Tokugawa homeland. He then became a newspaper editor and politician active in the Progressive Party and a supporter of Ōkuma Shigenobu. He used the twenty-seventh anniversary of Ii Naosuke's death to rectify Ii's vilification in the eyes of the pro-imperial historical narratives.[41] Critics of the Tokugawa shogunate often charged Ii with hastily signing treaties with the United States, intervening in the Tokugawa shogunal succession dispute, and initiating the Ansei Purge, which led to the execution of pro-Meiji men like Yoshida Shōin and Hashimoto Sanai. Even though Ii shared with others a dislike of foreigners, he put aside his personal feelings, Shimada argued, for the sake of the nation.[42] Shimada's apologetics regarding Ii foreshadowed many themes found in Oguri's subsequent rehabilitation: great foresight ahead of his time and thus misunderstood, a true patriot, a creator of the foundation for Meiji Japan, a victim of lesser men's weaknesses.

Significantly, Shimada's was one of the first pro-shogunate works to appear during the Meiji Restoration boom of the 1890s. Despite restrictions placed on the press during the 1880s, as the Popular Rights Movement collapsed, political commentary was taken up by authors of political novels.[43] Government critics increasingly called for a "second Restoration" during the 1880s to complete what they believed was an unfinished Meiji Restoration; their calls coincided with Restoration commemorative events, such as Ii Naosuke's anniversary and the rise in Saigō's popularity following his pardon in 1889. These critiques of the government invoking the Restoration continued into the 1890s, a ripe time for publishing, as Marvin Marcus has shown. There was rapid development of mass literary journalism, exemplified by the Min'yūsha, as well as a boom in biographical works, including the first "modern" biography (about Saigō Takamori) in 1895.[44]

Oguri's image benefited from this Restoration boom. Ex-retainers, led by Oguri's colleagues, established groups to support Tokugawa commemoration: the *Kyūkōkai* was one of the first, followed by the *Hekketsukai*, composed of veterans who fought in Hakodate; the *Hassakukai*, which planned the three hundredth anniversary of Edo; and the *Dōhōkai* and *Edokai*, each with its own commemorative publications. Not surprisingly, many ex-retainers belonged to more than one group; Enomoto Takeaki, Kurimoto Joun, Kimura Kaishū, and Katsu Kaishū formed the core leadership of several groups. Oguri is mentioned in many of their early pro-Tokugawa books and within magazines such as *The Nation's Friend (Kokumin no tomo)*, the *Journal of the Historical Association of Japan (Shigakkai zasshi)*, and *The Former Shogunate (Kyūbakufu)*. Of these, *The Former Shogunate*, published from 1897 to 1900, gathered together the most materials relating to the Tokugawa shogunate, including memoirs of surviving shogunate officials. Men behind the journal, such as Katsu, Kimura, Kurimoto, and Togawa Zanka, also published their own books rehabilitating and explaining the final years of the shogunate, supporting one another's efforts to rescue the Tokugawa legacy from Meiji history writing. Shimada Saburō and Taguchi Ukichi, for example, contributed the preface to Kurimoto Joun's 1892 *Ten Essays by Hōan (Hōan jisshū)*.[45] In it, Taguchi praises Oguri for his efforts in advocating the opening of the country to the West, while Shimada refers to "Oguri and others" for their involvement with the Yokosuka shipyard.[46] In 1896, Shimada used Kurimoto's story of "a house with a storehouse attached" as proof that Oguri was not simply working for the shogunate but, by supporting the shogunate, working for Japan's future.[47]

Some praised Oguri because they knew and respected him. For example, Kimura Kaishū, Oguri's former subordinate on the 1860 embassy, listed Oguri among the eighty-eight important shogunal officials. Kimura's work was a reaction against the popular "civilization" view of history (*bunmeishi*) that downplayed roles of individuals in history and emphasized a macro-natural progression of history over time: "Although one could say that the ups and downs of a nation arise from natural momentum, most states are affected by the loyalty and qualities of their vassals."[48] In Kimura's view, Oguri demonstrated this with his proposed military and economic reforms, his role in the 1860 mission, and the building of Yokosuka.[49]

Many who appropriated Oguri situated him within a larger scheme of historical verities that questioned how Oguri and others were treated by the Satchō forces. Togawa Zanka, editor of *The Former Shogunate*, wrote a short, two-part Oguri biography featured in two of the journal's later

issues. Although he added nothing new to Oguri's biographical information, relying on earlier works, Togawa located Oguri's life and death
within the journal's larger polemic. Whereas others similarly bemoaned
Oguri's execution and only briefly, and indirectly, criticized the Satchō
leadership for executing him, Togawa prefaced his Oguri biography by
invoking the arbitrariness of history: "Yesterday's enemy of the court
becomes today's royal servant, last year's servant is this year's enemy of
the court. . . . Ah, who determines the standard? That which is called
righteousness or unrighteousness, sincerity, or dishonesty is determined by strength."[50] This, for Togawa, explained why Oguri had been
ignored in contemporary history writing.

Togawa's description of Oguri derived from a composite of features of
other historical figures. In Togawa's view, the two most progressive
statesmen in the *bakumatsu* period were Ii Naosuke and Mizuno Tada-
kuni. Oguri's personality made him similar to these men: strong, resolute,
and forthright; had Oguri been born a vassal lord, he would have ruled
brilliantly, like Ii.[51] Although Togawa acknowledged Oguri's importance
as a military reformer and diplomat, he believed that it was his financial genius that placed him above his contemporaries: "Before Oguri came
along neither the senior councilors nor the financial commissioners
understood economy, they were simply rubber stamping whatever was
placed in front of them. The government wasted money, only Oguri knew
how to use it."[52] Togawa, like others, praised Oguri for his work building
Yokosuka and establishing Hyōgo Shōsha, blaming "the times" and the
Tokugawa regime's weakness for Oguri's failures. Togawa furthermore
placed Oguri within the familiar trope of other tragic heroes, namely
Yoshitsune and Saigō Takamori, who avoided unjustified, violent deaths
at the hands of the state, escaping Japan to return again: "There are even
rumors," Togawa mused, "that the imperial army did not kill Oguri, but
that he went abroad and will return to Japan and take the post of Minister
of Finance."[53]

Togawa appropriated Oguri not only as a way to lash out against the
Satchō-led Meiji government but also to attack fellow contributors to *The
Former Shogunate*. Togawa had been a member of the Shōgitai, a pro-
Tokugawa militia that fought the Satchō-led coalition. Oguri's connection to the Shōgitai is unclear, though he did have contact with some of
its senior members in 1868. In death, however, Oguri and the Shōgitai
fallen were part of the same tragedy: "Aa! May the spirits of those brave
Shōgitai members who died in battle welcome Oguri at the gates of
yomi."[54] Oguri, in Togawa's view, was a tragic figure not simply because
the Meiji government had killed him arbitrarily but because lesser

Tokugawa retainers became Meiji heroes despite Oguri's greater contributions to modern Japan. "In Oguri's day, men like Katsu Kaishū and Enomoto Takeaki had nothing to do with the shogunate's operation [as Oguri did]. It's true what they say—what happens in the world is a matter of luck."[55]

To Togawa and others, men like Katsu and Enomoto had betrayed the Tokugawa by eventually siding with and working for the Meiji government. Even Fukuzawa Yukichi lambasted those who took positions within a government dominated by their enemy.[56] Togawa's attack on Enomoto and Katsu, in fact, alluded to a short essay written by Fukuzawa in 1891. Called *On Fighting to the Bitter End* (*Yasegaman no setsu*), it is a work that receives little attention in Western-language scholarship on Fukuzawa. Fukuzawa, indeed, never intended to publish it, and showed it only to Kurimoto Joun, Kimura Kaishū, Enomoto, and Katsu, among a few others.[57]

Fukuzawa's personal connections to these men informed his political motivations for writing the essay and showing it only to a select group. The rivalry between Katsu and Fukuzawa can be traced back to the 1860 embassy. Fukuzawa's benefactor, Kimura Kaishū, acted as the diplomatic leader of the *Kanrin Maru*, while Katsu captained the ship. Katsu was the older man, and he loathed taking orders from Kimura.[58] This animosity passed on to Fukuzawa, who claimed to respect Katsu but was quick to criticize him. For example, in his autobiography, he called Katsu a poor sailor who never left his cabin on the *Kanrin Maru*, and poked fun at him for reneging on a vow to hand over his head if the sailors disobeyed him by shooting a salutation as they arrived in San Francisco.[59] Fukuzawa, who loathed writing forewords to others' books, gladly wrote one for his mentor's *Thirty Year History* (*Sanjūnenshi*), noting Kimura's success on the *Kanrin Maru*.[60] Kurimoto did not get along well with Katsu either, as Katsu explained it, "At first, Oguri really favored me. But since the incident of borrowing money [from the French], and this includes Kurimoto, that whole gang started treating me like an enemy. Kurimoto never came to see me after that."[61]

In the essay, Fukuzawa seeks to explain Japan's native spirit, informed by individual emotions and exemplified by the Mikawa samurai, the defeated bannermen retainers with long-standing connections to the Tokugawa. This spirit acted as the defining idea in the forming of the nation. Countries and subregions form naturally from a shared identity, and some will fight to protect their group at all costs. Therefore, a strong country can only be maintained by holding on to this private emotion of fighting to the bitter end. Fukuzawa appropriates Oguri's words: "One extreme example is the reaction of people when their parents are stricken with a terrible sickness. They will hope for their recovery and will not stop trying to find a cure until the actual moment of death. Wise people give morphine to make death easier for the sick, but when the sick person is a parent they'll do anything to extend life by even one day, holding on to hope."[62] A strong country can only thrive if the bond between the state and its subjects' emotions are as strong as that of parent and child.

Though Fukuzawa does not lament the shogunate's collapse—he criticized its feudal and backwards nature in other writings—he nonetheless contends in this essay that the "beautiful spirit" of the Mikawa samurai, that desire to fight to the end knowing that they would lose, should not be discarded. Fukuzawa acknowledges that Katsu was an impressive man whose submission to the Meiji forces brought peace and saved lives, but in so doing, he argues, Katsu abandoned the Japanese martial spirit, embarrassing Japan internationally and leaving shame for future generations. Even worse, Katsu happily sided with the new government, put on airs, became wealthy, and was considered by many to be a hero.[63] Fukuzawa similarly criticizes Enomoto, because although he resisted the new government, he also ultimately joined it and therefore insulted the memory of those who died fighting with him at Hakodate.

Fukuzawa's piece immediately became the most talked about essay of the time.[64] Katsu's and Enomoto's responses were published: Enomoto replied only that he was busy and would take a look at the essay when he had time, while Katsu stated that he took responsibility for his actions, and it was up to others to judge him.[65] The journalist, writer, and activist Tokutomi Sohō, a generation younger than Fukuzawa and Katsu, published a rejoinder in *Kokumin shinbun* a few weeks later.[66] He criticized Fukuzawa's essay for being nothing more than an academic exercise, just a "snake weaving through the grass."[67] Tokutomi agreed that one should admire the guts shown by men like Oguri, but if the shogun had chosen to fight, then the French, English, and even Russians would have intervened. According to Tokutomi, this was what worried Katsu, and thus his political maneuvers should be praised.[68] Tokutomi's own views on history and politics changed broadly over time, from advocating for the democratic rights of the people (*heimin shugi*) to supporting imperial expansion.[69] In his early career, when he published *The Future Japan* (*Shōrai no Nihon*), Tokutomi decried any hint of militancy, which he saw as an impediment to social progress.[70] This might explain his preference for Katsu's peaceful resolution to the Restoration over Oguri's militarism. It is true that by the early 1890s, and increasingly so after the Sino-Japanese War and Triple Intervention, Tokutomi gradually began to support Japan's colonial and military expansion abroad, which suggests that he would come to support a military figure like Oguri. But at the turn of the twentieth century, Tokutomi chose political pragmatism over popular politics and distanced himself from liberal idealism while forming closer ties to the oligarchy. Tokutomi's sympathies, then, lay with Katsu, a similarly maligned political pragmatist. Again, personal ties also mattered. Katsu was a neighbor, landlord, and benefactor who comforted Tokutomi when popular opinion turned against him as he shifted from government critic to supporter, a change that also affected his livelihood as a journalist.[71]

Fukuzawa's strongest supporter and former newspaper employee, Ishikawa Mikiaki, published a counter essay that accused Tokutomi of not knowing the facts surrounding the shogunate's diplomatic situation. Ishikawa focused his argument around Oguri, whom he called an exemplary Mikawa retainer with a well-known family connection to Tokugawa Ieyasu.[72] By highlighting Oguri's interaction with Roches, Ishikawa illustrated just how clever these Mikawa samurai could be: Oguri introduced the Chinese medicine master Asada Sōhaku to Roches, and used Roches's recuperative vacations to Atami as a pretext for sending Asada to act as messenger. Ishikawa noted appreciatively "now that shows some diplomatic skill."[73]

The struggle over the historical narrative evident in the putative rivalry between Oguri and Katsu, a major theme among Meiji-period writings by former Tokugawa men, has continued to the present. Itō Chiyū summarized nicely how the two face each other in the historical memory: "People who believe Oguri was great will find much at fault with Katsu, and those who sympathize with Katsu's stance will think Oguri was short-sighted."[74]

For his part, Katsu, in his writings, was generally critical of most Tokugawa men.[75] He respected Oguri's spirit as a Mikawa samurai but disagreed with his views on preserving the shogunate. As Katsu saw it, Oguri had risked Japan's sovereignty by depending on the French. For Katsu, only Saigō was deserving of high praise: "There are lots of biographies about people from the Meiji Restoration period, but the only person of any value to talk about is Saigō."[76] Not coincidentally, Katsu co-wrote such a biography, and he frequently mentioned Saigō in his memoirs.[77] Katsu's own place in history is validated by his legendary meeting with Saigō when Katsu mediated the last shogun's surrender to the emperor's troops.

Saigō's legacy seemed to be the only point of agreement between Fukuzawa and Katsu. Fukuzawa's *On Fighting to the Bitter End* was serialized in 1901 and was eventually published together in book form with another controversial essay titled *Commentary on the Problems of the Year 1877 (Meiji jūnen teichū kōron)*. Fukuzawa claimed to have written the essay to defend Saigō's actions at a time when many commentators thought negatively of Saigō, but did not publish it because he feared breaking the Libel Law (1875).[78] Here too, Fukuzawa argued that although he did not advocate violence, he admired Saigō's spirit of resistance against a corrupt government. As in his essay *On Fighting To the Bitter End*, Fukuzawa used vilified historical figures to argue that this spirit of resistance was not just about individuals fighting against a tyrannical government but about an attitude of independence that Japan needed to resist other countries as well.

One can see Fukuzawa's sympathies for the ex-Tokugawa retainers even in his essay about Saigō. He lamented: "[The] tens of thousands of retainers who moved to Shizuoka and passed away in obscurity, or begged for food in the alleyways of Tokyo, had their homes confiscated while the imperial officials lived a life of ease, were left in cemeteries to deteriorate and become the homes of foxes and badgers; it was a pitiful scene, one that I could not bear to see."[79] These tough conditions existed for most of the ex-samurai, not just retainers, and Fukuzawa admitted that Saigō was a major figure responsible for these changes in fortune. Ultimately, however, the advance of civilization and the improvement of conditions for the people were good. What angered Fukuzawa was the treatment of men like Saigō and Etō Shinpei, another rebel leader, despite their good works. Just as Oguri supporters pointed out that Oguri was killed without having a trial while combatants like Enomoto were pardoned, Fukuzawa complained, "the crimes against the state are to be hated, and those who committed them also hated, but sometimes these are forgiven. . . . Enomoto and others were pardoned and today do not suffer at all. During the Saga Rebellion too, Etō was arrested and killed without a public trial. This is not punishment, but simply murder on the battlefield."[80]

While Fukuzawa admired the shogunate's loyal supporters rather than the shogunate itself, Fukuchi Gen'ichirō began a rehabilitation of the larger Tokugawa legacy. His post-Restoration historical writings sought to counterbalance those versions of Meiji Restoration history that favored the victors and force the public to consider the value of the Tokugawa shogunate and its leaders. This extended to the Tokugawa family itself. Despite the Meiji emperor's formal recognition of Yoshinobu's support for the imperial line, others were not as forgiving. At the Gakushuin school for nobility, some Tokugawa descendants skipped lessons on

the Tokugawa period to avoid embarrassment, and one Tokugawa girl
cried when the teacher called Ieyasu a "sly old badger" (*tanuki
oyaji*).[81]

By the time Fukuchi began publishing these histories, he had already expanded his career from a Restoration-era newspaper pioneer,
who covered the fighting and deaths of Tokugawa supporters, to a modern journalist, essayist, playwright, and politician. After working under
Itō Hirobumi in the Conference of Local Officials in 1878, he ran against
Fukuzawa for the president's position on the Tokyo City Council in 1879.
He won, and Fukuzawa, perhaps a sore loser, resigned the second place
vice president's seat due to the "press of business."[82] Fukuchi later created
a political party, the Constitutional Imperial Party, though it never
gained popularity. He became a Diet member in 1903 for the last few
years of his life, after his career as a writer and journalist had plateaued.

Fukuchi's two apologetic works on the Tokugawa shogunate offer a
good illustration of the "Meiji bias" and, like most other writings by
Tokugawa men, are mostly memoir centered. Given his early pro-
Tokugawa news reporting, his own ties to the shogunate, and his prolific
writing outside of the journalistic genre, it is no surprise that as a historian, he would write a history of the Restoration period from an insider's
perspective. The pro–Ii Naosuke works of the late 1880s also influenced
Fukuchi's decision to write a history of the fall of the shogunate.[83] Politics alone does not explain Fukuchi's support for the Tokugawa legacy.
He had no simplistic hatred of the Meiji oligarchs; in fact, one scholar
has suggested that Fukuchi's political aspirations and connection to the
oligarchs tempered his criticism of the government, and thus his writings about Oguri also shifted.[84]

Fukuchi published his first and most influential work, *On the Decline
and Fall of the Shogunate* (*Bakufu suibōron*), in 1892, after it had appeared
serialized in Tokutomi's journal, *The Nation's Friend*. Fukuchi argues
that the Tokugawa shoguns did not usurp imperial power; instead, their
feudal and isolationist policies created stability in Japan. Far from being
anti-emperor, the shogunate fostered loyalty to the emperor by invoking
his name in various administrative and policy rituals that led to the Restoration.[85] The Satchō leaders never restored the emperor's rule but simply created another shogunate.[86] Accounts of nobles and high-ranking
daimyo and shogunate policy makers comprise the bulk of his analysis,
while he notes that low-ranking warriors, men like himself and Katsu,
did not affect much—a subtle poke at the Meiji oligarchs who were mostly
men of low rank.[87] Oguri is mentioned alongside fellow retainer Iwase
Tadanari, who expressed frustration with the lack of preparation against

the shogunate's enemies. According to Fukuchi, Oguri stated that "if there was one phrase that would cause the country to fall, it would be 'well, what're we going to do.' That would destroy the shogunate."[88] Likewise, Iwase is supposed to have said, "There are three words that should be forbidden during a shogunate cabinet meeting, '*naru beku take*' (what should happen will happen)."[89] These comments, Fukuchi argues, indicate the indecisiveness that Iwase and Oguri were trying to fight against within the shogunate.

Although Fukuchi's book only mentioned Oguri in passing, it laid the foundation for his treatment of Oguri in *Statesmen of the Bakumatsu Period (Bakumatsu seijika)*.[90] Most of the work covers the contributions made by three senior members of the shogunate in its last years: Abe Masahiro, Hotta Masayoshi, and Ii Naosuke. In a short, final section, however, Fukuchi devotes attention to three lower-ranked functionaries involved with foreign affairs, all of them men with whom he had interacted: Iwase Tadanari, Mizuno Tadanori, and Oguri.[91] Fukuchi believed that of the many Tokugawa retainers, these three were "the three great men of the *bakumatsu*, and this is not an overstatement. This is not just my praise, but is a feeling shared by Kurimoto Joun and Asahina Kansui (Masahiro)."[92] Fukuchi argued that despite their devotion to the shogunate, Iwase, Mizuno, and Oguri were never promoted to higher ranks and could never demonstrate the extent of their talents. Moreover, Iwase and Mizuno died filled with resentment, and Oguri was killed without having committed a crime. Fukuchi mourns them in a poem: "Ah, heaven, is this fair or foul?"[93]

Fukuchi's evaluation of Oguri's life and death established a precedent for subsequent Oguri commentaries. Fukuchi claimed that Oguri himself stated, upon returning from the 1860 embassy, "We must improve our country by using foreign countries as a model in the fields of government, military preparedness, commerce, and manufacturing."[94] To a well-educated Meiji readership, Oguri's strategy fit into the Meiji government's policy of building a "rich country, strong army." Meiji statesmen believed that the foundation of Western strength lay in its naval capabilities and its industrial and commercial infrastructure, a point driven home during the 1872 Iwakura embassy. Fukuchi and other former shogunal officials also drew attention to the Tokugawa regime having initiated many of the Meiji state building projects, modeled on Western examples. He noted that Oguri imported goods from France and Great Britain in addition to hiring French instructors to drill the army, establishing a military school, and building Yokosuka. The Meiji government mimicked these policies, such as hiring foreign workers in a variety of industries, from railroad building to beer brewing. He deemed Oguri's contribution to commerce important as well. Although Oguri could not convince merchants to support Hyōgo Shōsha, Fukuchi blames senior shogunal men for this failing, because they argued against the plan by asserting that the time was not right for such a venture. For Fukuchi, Oguri was a hero for refusing to ignore the shogunate's problems and never abandoning difficult projects.[95]

According to Fukuchi, Oguri's stubborn loyalty was both the source of his downfall and the reason why readers should remember him. He explained the contradiction between Oguri's loyalty to the shogunate and farsighted reforms in the following manner: "It is not filial to deny parents their medicine just because you know their sickness is incurable; likewise, a true *bushi* works for the public until the country falls apart."[96] In Fukuchi's view, Oguri appears not as a Tokugawa absolutist, an accusation often made against him, but as a patriot.

Whereas Fukuchi made a particular effort to champion Oguri, the first full treatment of Oguri's biography was written not by a nationally recognized writer but by Tsukagoshi Yoshitarō (Teishun), a local intellectual from a village near Gonda. Tsukagoshi's career demonstrates the increasing role that intellectuals from rural areas outside of Tokyo could play in the formation of the nation, in his case, in shaping national historical views. Tsukagoshi was born in Iwakōri Village in 1864, part of the five-village community that included Gonda and Mizunuma, where Oguri was killed. In his youth, Tsukagoshi studied in a Chinese school established by Satō Tōshichi, the Gonda Village headman, and, as an adult, became the principal of several local elementary schools. His involvement with the movement to abolish licensed prostitution brought him to Tokyo, where he met fellow abolitionist Tokutomi Sohō and started working for Tokutomi's Min'yūsha company in 1890.[97] Tsukagoshi's background mirrored that of other Min'yūsha intellectuals, men like Taguchi Ukichi, Shimada Saburō, and Yano Fumio: educated, wealthy, socially elite men from the countryside who believed that Japan's future leadership lay in the hands of "country gentlemen" such as themselves.[98] They tended to view the Meiji Restoration as a move from feudal warrior Japan to industrial Japan.[99] Tsukagoshi's biography of Oguri, which appeared in *The Nation's Friend* in 1893, depicted him as the first Tokugawa retainer to push this transition through modernization efforts exemplified by Yokosuka and Hyōgo Shōsha.

Tsukagoshi's extensive biography of Oguri began with an evocation of sadness, death, and the local setting. "There is a place," Tsukagoshi wrote, "where the old people in the village say the great *bakumatsu* politician Oguri Kōzukenosuke hoped to build a house . . . but all you can see now are fallen leaves."[100] Oguri's hopes for the future failed, and the graves of his retainers and successor lie next to his behind the Tōzenji Temple, their grudge for having been falsely persecuted knowing no end. To express his grief, Tsukagoshi quoted a poem by Fujiwara Shunzei. After visiting his mother's grave, Fujiwara wrote:

I come so seldom,	*Mare ni kuru*
and yet how sad in the night	*yowa mo kanashiki*
sounds the wind in the pines.	*Matsukaze wo*
And she, there beneath the moss—	*taezu ya koko no*
does she too hear it, endlessly?	*shita ni kikuramu*[101]

Beginning his narrative in this way, Tsukagoshi momentarily removes Oguri from the context through which he is typically viewed—his life in Edo—and places him in the rural setting where he spent his last days. Oguri's death, moreover, automatically kept him from being categorized as one of the contemptible "old men of Tenpō," a phrase used by Min'yūsha intellectuals to refer to the representatives of old Japan, specifically the Meiji oligarchs whose policies intellectuals wanted to critique.

Tsukagoshi rescues Oguri from the margins of Tokugawa history by locating his policies within the framework of "civilization." Oguri's desire to create a postal system, produce government-issued paper currency, and build Yokosuka exemplified his deeds as a politician within a civilized nation.[102] It did not matter that Hyōgo Shōsha failed; Oguri's effort to issue government loans to merchants in exchange for their investment demonstrated not only his deep economic knowledge but also the value he placed in the idea of trust.[103] Although shogunal officials hated Oguri for borrowing money from the French, his forward thinking marked him as a "constructive progressive" (*kensetsuteki kaikakuka*).[104] For many, this project and his efforts to destroy the Satsuma and Chōshū domains demonstrated that Oguri was a Tokugawa absolutist. For Tsukagoshi, however, these were signs that Oguri's ultimate goal was to improve the nation:

> Sure, Oguri was the strongest supporter of the shogunate, and he amassed foreign debt to defeat Satsuma and Chōshū, but he was nothing less than a patriot of the times. Unlike the small, close-minded patriots who, focusing their loyalties on the court, advocated expelling the barbarians, Oguri was a patriot who worked for the long-term good of Japan. . . . He did what he could within his position, and understood the meaning of "civilized patriotism (*bunmeiteki aikoku*)."[105]

Here Oguri became the ideal hero within the confines of the Min'yūsha intellectuals' Whig narrative of progress and civilization. In this version of Oguri's biography, he is portrayed as a visionary uninhibited by political considerations who acted for the good of the country, thus undercutting the putative achievements of the Meiji leadership.

Tsukagoshi's work directly influenced later writings about Oguri. Many subsequent biographers accepted, uncritically, Tsukagoshi's details and interpretation of Oguri, sometimes lifting them word for word. For example, Kawasaki Shizan's short biography of Oguri and his wife Michiko in *History of the Boshin War* (*Boshin senshi*), published in twelve volumes from 1893 to 1894, copied Tsukagoshi's work verbatim. Kawasaki even borrowed Tsukagoshi's portrayal of Oguri's character: "Oguri did not partake in Noh, nor did he compose poetry or drink wine. The one thing he enjoyed was a little bit of writing. He did not collect antiques nor was he impressed by the ability to distinguish items from the Song or the Ming."[106] The message was clear to a Meiji-period audience who understood the government's educational emphasis on practical skills: Oguri's contemporaries pursued extravagant and wasteful lifestyles, but Oguri exemplified the ideal of living a simple life.

Oguri's importance as an exemplar of values among other Restoration losers was especially evident in the only Meiji-period book to appear about him: a young person's reader (*shonen dokuhon*), written by Seta Tōyō and published by Hakubunkan in 1901.[107] During the 1890s, Hakubunkan succeeded in publishing more biographies than its competitor, Min'yūsha, and, in 1899, created a children's biography series reflecting the genre's value as a pedagogical tool.[108] Yamaguchi Masao notes that because Hakubunkan's founder, Ōhashi Sahei, was born in the pro-shogunate Nagaoka domain, many of its publications tended to favor themes about the losing side of the Restoration.[109] Other historical figures in the series include Kawai Tsugunosuke (written by Togawa Zanka), a Nagaoka samurai who fought as a military commander against the Satchō forces during the Boshin War; Takashima Shūhan, a drill instructor and gun importer for the shogunate; and Ii Naosuke. Anti-shogunate figures also appear, such as Saigō Takamori and Sanjō Sanetomi, but they tended to be more well-known, whereas the shogunate men were more obscure.

Seta's book includes long excerpts from Edo-period documents on a variety of Oguri-related events. He often cites early Meiji period writings about Oguri, including works by Katsu Kaishū, Kurimoto, and Fukuchi, and, as did other Meiji writers, adopts some of the exact phrases found in Tsukagoshi's articles. However, Seta's audience differed from that of his predecessors in that he wrote for children. After a brief narrative of Oguri's contributions to Japan, Seta quickly arrives at his point: that Oguri was stubborn, resolute, and able. Although these qualities were a source of his strength, they also earned him the hatred of his superiors, who often held him back. "In this world there are politicians who try to be all things to all people but Oguri stayed his own course."[110] Oguri's superiors disliked his attitude, but fellow bannermen also hated him, for cutting their salaries, and those who advocated "expelling the barbarian" attacked Oguri for his love of the West. No group escaped blame in Seta's work. The popularity of the anti-foreign faction was responsible for halting all progress, while the shogunate was depicted as incapable of sustaining progress.[111]

In Seta's view, Oguri was quite literally the cure-all for the shogunate's financial problems. He compared the shogunate's financial and economic ills to the difference between traditional Chinese medicine and modern Western medicine: "The public treasury is the heart of the state," quoting, in English, the Irish surgeon Robert Adams, "and the shogunate's heart stopped. . . . Oguri was the cure for the shogunate's heart failure."[112] Seta extended his analogy to the contemporary medical world, stating that Oguri provided the cure in the form of "Western medicine," like the Western-trained physician Satō Susumu. The anachronistic Chinese medicine—"the way of Asada Sōhaku," a Chinese physician in the modern Meiji court—was backwards and inefficient.[113] Typically, the Tokugawa period is equated with traditional Chinese learning, while the Meiji period is characterized by the dominance of Western technology (in this example, medicine). Perhaps unbeknownst to Seta, Ishikawa had previously praised Oguri's connection to Asada, but nevertheless, by reversing the familiar characterizations of the Tokugawa as traditional and the Meiji period as modern, Seta opened up the possibility that Tokugawa Japan possessed elements of progress just as Meiji Japan had its moments of backwardness.

Writers in Tokyo dominated the textually mediated memory about Oguri at the "national level," from his earliest appearances in Restoration newspapers to Seta's youth reader in the early twentieth century, but parallel local developments added to his legacy. Tsukagoshi's treatment of Oguri from a uniquely regional perspective was followed during the 1890s by Oguri's incorporation into specifically regional histories, suggesting that a small core group of village elites did, in fact, believe that they too benefited from their association with Oguri. Local history, memory, and identity formation was a local affair, one that remained important to forming a national identity, as Kären Wigen has recently argued;

regional identity and institutions were important to creating modern Japan.[114] Just as Shinano/Nagano place names, geography, topography, and local history created putatively unique regional characteristics that connected Nagano to Japan, so too did Gunma historians want to connect their histories to a greater, national whole. Oguri, as a national historical figure, became a convenient tool for accomplishing this.

The locally published 1893 *Biographies of Early Modern Jōmō Greats* (*Kinsei Jōmō ijin den*) followed the Oguri narrative already written elsewhere: his stoic ways, contribution to Yokosuka, constant switching of positions (an unlikely seventy times), and involvement in the incident with the shogun's *hakama*.[115] The *Usui County History* (*Usuigun-shi*) records Oguri's fight against bandits in the region and recalls the Yoshii soldier who helped Oguri's wife escape to Aizu.[116] In Gunma as well, Oguri started appearing in similar pedagogical contexts. A teaching supplement published in 1894, *Kōzuke Province History for Elementary School* (*Shōgaku Kōzukeshi*), offers information on local history and geography used by teachers in Gunma Prefecture to add to their lessons to promote local and national identity. In the book, the author, Iwagami Masai, highlights Oguri's loyalty to the Tokugawa cause, knowledge of the West, and vigorous work ethic. Iwagami believed that Oguri foresaw the collapse of the shogunate after its loss at the battle of Toba-Fushimi and thus had retired to a life of agriculture in Gonda. But unlike other biographers, Iwagami did not blame the Meiji government for Oguri's death, a controversy too sensitive for classroom instruction. He instead accused the rioters of falsely portraying Oguri as a bad person to the imperial court; it was only then that soldiers were sent to arrest and execute him.[117]

It is also at this time that we see the first commingling of regional identity with the growth of pro-Tokugawa commemoration in national writings by those living in Tokyo. Ōtori Keisuke, Oguri's former subordinate, a Meiji government official, and an editor of *The Former Shogunate*, visited Gunma in 1895 and promised locals that he would help in Oguri's commemoration any way he could.[118] Recalling his Gunma visit, Ōtori reassures his readers: "Since Oguri's death I had been quite worried about his legacy. Fortunately, I was able to visit the governor of Gunma, Nakamura Moto'o, last summer in Maebashi. When I asked him [about Oguri], he told me not to worry, that people there erected a grave marker and someone with a close connection to Oguri became the monk at his family temple [the Tōzenji]. . . . Satisfied and with peace of mind, I returned to Tokyo without having to go there myself."[119] These interactions between local and national elites who wanted to protect and appropriate Oguri's memory increased as the generation of those who knew him died.

Part Two: Oguri and Legend in Post-Restoration Kōzuke

Despite his brief appearance in local histories of the 1890s, Oguri's legacy in the countryside was largely confined to the realm of rumor, superstition, and legend. As Noguchi Takehiko argues, Oguri's supporters in rural Gunma and Saitama Prefectures used these legends to rehabilitate his name in the realm of "folk customs" (*dozoku*) because mainstream nineteenth- and early twentieth-century history had marginalized him.[120] Noguchi's understanding of the legend echoes Assmann and Czaplicka's idea of communicative memory—a liminal memory that survived, initially, outside of textually mediated memory. For communicative memory, the time horizon is short, lasting only a few generations. Even after WWII, local villagers could tell stories of their parents or grandparents encountering Oguri, his retainers, or the fighting in which he was involved. Gradually these rumors turned into tales of buried treasure or were spun in the direction of local history.

In 1871, the Meiji government split Kōzuke into a mix of small prefectures, most corresponding with former domains, before eventually creating what is known today as Gunma Prefecture. Among these, Iwahana Prefecture incorporated former shogunate and bannermen lands, including Gonda. Just as former Tokugawa men wrote from their position in Tokyo, within the context of the Tokugawa-Meiji transition and the burgeoning nation state, local uprisings and harshly enforced law under the Iwahana governor affected perceptions of Oguri's legacy in post-Restoration Kōzuke Province.

The narrative of marginality that facilitated Oguri's inclusion in the larger context of Gunma Prefecture history arose out of the traumatic Tokugawa-Meiji transition experienced in Kōzuke. As Gunma historian Nakajima Akira has shown, the local Meiji authority labeled Kōzuke people "enemies of the court" due to the frequency and magnitude of their uprisings.[121] They did not simply resist the new government but attempted to negotiate the changes they faced during this transitional period. In so doing, Nakajima argues, the Meiji government stigmatized Kōzuke (and later Gunma) citizens as vagabonds, outlaws, and poor peasants, marginalizing them in the historical memory of the Meiji Restoration.[122]

The transition pains were rooted in the complex land arrangement in the Kantō region, which posed a unique challenge to Meiji leaders. The government needed to check potential enemies among samurai who lived in the region and whose sympathy lay with the shogunate: vassal daimyo and their retainers, former Tokugawa bannermen like Oguri, and others who administered land owned by the former shogunate. Moreover, the

gap between the end of the Tokugawa regime and the salient beginning
of a new government allowed commoners to assert themselves against
local authorities. Active and passive resistance against the Kantō Regulatory Patrol continued even as the new Meiji leaders created Iwahana Prefecture.[123] The Tōsandō army took over the former Iwahana intendant
station, replaced its symbolic Tokugawa authority with a Meiji one, and
installed a military government under the harsh rule of its "temporary
governor and military director" Ooto Ryōtarō. Ooto, as the chief inspectorate for the Kantō region, supervised Oguri's persecution and, as an
unpopular governor, acted as a flashpoint for a tense relationship between
the early Meiji government and local citizens.[124] Although local historians portray Ooto's tenure as a "government of fear," his rule over the haphazardly created Iwahana Prefecture illustrated the problems facing the
early Meiji leaders.[125] The sudden land conglomeration is thought to have
been the main cause of the subsequent uprisings throughout Kantō that
impelled the newly formed Council of State to appoint military men to
prefecture governor posts.[126] The Meiji government wanted Ooto to quell
the peasant uprisings that plagued Kōzuke and challenged the establishment of imperial rule. Local bureaucrats carried out the prefecture's day-
to-day affairs, but Ooto set the agenda. He announced to village officials,
"The imperial court has appointed me general and governor of this prefecture. I will praise good citizens, punish the bad ones, and look after
the poor. I will make each citizen understand the appreciation owed
to the court. To achieve this, you will obey my command as if it were a
direct order from the court."[127]

A much abler persecutor than administrator, Ooto became known
as "Ryōtarō the People-Cutter," having once rounded up and executed
three hundred bandits.[128] This was more than just rumor; one Iwahana
official recorded in his diary the successive beheadings that Ooto ordered.[129] Ooto carried out his duties with zeal, as a former imperial soldier noted: "On the 24th Toyonaga and Hara fought in different domains
throughout Jōshū [Kōzuke], and attacked the Aizu rebels at the Mikawa
Pass. The rebels abandoned their weapons and lost. The heads of three
rebels were posted at Nagai Station. When Ooto heard this he was over-
joyed and clapped his hands."[130] On 10/1868, a representative of the village officials petitioned the court to have Ooto dismissed, complaining
that Ooto accepted bribes, punished citizens arbitrarily, and caused general suffering among the common people. The petition's language invoked government benevolence, a common strategy among Tokugawa
peasants as well: "We common people know that the imperial government's renewal is meant to stop evil and encourage good" and "the impe-

rial government does not do such things, but we believe Ryōtarō's actions are great crimes against the nation."[131]

Ooto's harsh but short-lived rule, plus local attempts to negotiate the post-1868 government transition, affected subsequent historical memory of Oguri. As the case of disorderly Iwahana Prefecture demonstrates, the early Meiji government failed to pacify Kōzuke completely. Although the government responded to complaints against Ooto by dismissing him, many citizens still resented the Meiji government's policies. Although the language in the petition to dismiss Ooto acknowledged Meiji legitimacy, the transition from early modern Kōzuke Province to modern Gunma Prefecture acts as a background of tension and disorder within which the new government wronged both Kōzuke people and Oguri. Even Ooto's actions did not fit into any neat category; shortly after Oguri's execution, he gave twenty-five *ryō* to the Gonda village headman to build Oguri's grave and hold memorial services.[132]

In addition to the stresses facing Iwahana citizens in general, residents of Gonda Village and the surrounding area and those who worked for Oguri had to contend with Oguri's vilification by the local government. Some villagers continued to support Oguri's family after his execution. They were not obligated to escort Oguri and Tsukamoto's family out of Gonda Village, and handing them over to the imperial government would have quelled any fears that the village could be accused of harboring anti-government sentiments. They chose to take risks in honor of Oguri, as two incidents demonstrate. The first was a daring effort to retrieve Oguri's decapitated head. Although the head's eventual whereabouts are unclear, documentary evidence proves that several of his retainers stole it from the temple grounds in Takasaki City and reburied it, most likely in the Tōzenji Temple grounds.[133] The second involves the leader of the Oguri family escort, Nakajima Sanzaemon, who returned to the village after taking the Oguri women to Aizu, Tokyo, and finally Shizuoka, where many former shogunate men and their families followed the last shogun into retirement. According to local historians, and probably based on oral history, Nakajima reported his involvement with the Oguri family to the Maebashi domain office that administered Gonda Village. Instead of arresting him, the Maebashi officials told Nakajima, "You are not Oguri's retainer, you are just a village official, nothing more than a peasant, but hereditary vassals do not approach what you did. Oguri Kōzukenosuke must be satisfied with your sincerity." They gave him the surname "Gonda" using the Chinese character for "honor."[134]

The motivations for support of Oguri among village elites can be explained within the context of the changing status system during the

Tokugawa-Meiji transition. Village retainers who fought alongside Oguri during the Gonda riots had been educated at his Edo residence and thus given an opportunity to move beyond the confines of their status. These young men fulfilled the samurai ideal, legitimately receiving martial training and participating in a higher cause. Oguri's Gonda retainers, it seems, similarly demonstrated an idealized samurai loyalty. The example of Nakajima's name change, even if the oral history was not "true," indicates that at least he and other peasants thought of themselves in terms of samurai values. Oguri's death did not eliminate the social and cultural prestige they had accrued from their relationship with him by defending his legacy; they were protecting their new status.

Not all of Oguri's associations in the village were so warm. Some people suffered from their connection to him. After Oguri's execution, the imperial army searched through the homes of village officials, looking for Oguri's possessions not already confiscated from the Tōzenji.[135] Boshin War records indicate that several Oguri retainers died fighting against the Meiji forces in Aizu.[136] Others, such as Nakajima Sanzaemon, continued to escort the Oguri family out of Aizu. Among those survivors, some stayed away from Gonda for over a year, fearing reprisal, and a few never returned. One example of the latter is Satō Ryūsaku, who, according to his descendent Satō Hisao, never returned to Gonda, fearing punishment by local authorities.[137] Finally, there was the mysterious circumstances surrounding the death of Oguri's Gonda confidant, the village headman Satō Tōshichi. Villagers today still pass on the story of Ikeda Denzaemon, an Oguri retainer from Gonda who accompanied the Oguri family to Aizu. After leaving Aizu, Ikeda moved to Kyoto, where he worked as a police officer. He suddenly appeared in Gonda again in 1872, wearing his police uniform, and visited Satō Tōshichi's residence, calling him "our master's enemy." Satō fled through a back entrance, and Ikeda's father-in-law intervened to stop Ikeda's assault. That year, Satō mysteriously died; rumors continue to this day that he either committed suicide or was poisoned.[138]

It is this last feature of rumor—suspicion of neighbors and the new government alike—that characterized much of the early, local communicative memory of Oguri. Hijikata Hisamoto, Imperial Household minister in the late nineteenth century, related a story from Toyonaga, co-leader of the group that arrested Oguri:

> After Oguri was executed, villagers believed that because Oguri and his son glared at the villagers at the moment of decapitation, something bad would happen. Even though Toyonaga tried to reason with them, they did not understand. So Toyonaga told them, "Since we beheaded Oguri and his son, just

write our names on paper and post them on your doors. Then your village will not be cursed." The villagers thanked him and followed his advice. There is probably a paper or two still around today.[139]

The villagers' fear illustrated their ambivalent relationship with Oguri. Toyonaga seems to have cleverly capitalized on this when, rather than dismissing village superstition, he invokes the supernatural to assure the villagers that the new imperial government had properly usurped Oguri's authority. It was superstition, rumor, and legend—the world of the fantastic prevalent during this uneasy time of transition—that defined Oguri's legacy differently in the countryside than among intellectuals in Tokyo.[140]

Legends of treasure, however, became more permanently attached to Oguri's legacy. From 1868 to the present, booms in treasure hunting related to Oguri have occurred during times of crisis in Japan, beginning with the Tokugawa-Meiji transition. The legends began in 1868, when local gangsters, hearing that Oguri's entourage carried boxes and jugs filled with gold coins, led a riot to steal his money. According to the Iwakōri Village headman, after Oguri's execution, government soldiers stayed in Gonda to search for money and, having found none, still took everything, "not even leaving clothing behind."[141] When the government sold off some of Oguri's possessions that had arrived by boat, a local merchant bought several items and found money hidden in the bottom of one of them, a miso jug.

The most well-reported hidden-treasure legend involves ten trillion yen worth of shogunate money supposedly buried on Mount Akagi, Gunma Prefecture. The legend originated with Mizuno Tomoyoshi, born in 1851 as the third son of a Tokugawa retainer. One day Mizuno received a mysterious letter from a former neighbor, Nakajima Kurando, who claimed to have worked in the finance commissioner's office. Some contend Nakajima had had an illicit relationship with Mizuno's mother, while others say he was Mizuno's uncle; either way, the Mizuno boys had always called him "step-dad."[142] Nakajima claimed that he had helped transport 240,000 *ryō* from Yamanashi Prefecture to Mount Haruna in Gunma during the first four months of 1868, with tens of thousands more *ryō* coming from Edo. However, when he later returned for the money, it was gone, and he believed it was reburied on Mount Akagi. According to testimony given during a 1933 court case, Nakajima gave an old Keichō gold coin as down payment (for undisclosed reasons) to an American employed on a railroad project in Yokohama, with the promise of more. Nakajima then dug on Mount Akagi, but when he could not produce more coins, the foreigner sued him. (In another version, Nakajima defrauded the American by selling him land and convincing him to dig for treasure on Akagi.) As a result, Nakajima spent two years in a Yokohama City jail. After his release, he lived off of money borrowed from Tomoyoshi, causing trouble for the Mizuno family. On his deathbed he told Tomoyoshi about the hidden treasure on Mount Akagi, near their homes. In 1888, Tomoyoshi used money earned as a Tokyo merchant to start digging at his family's home.[143] Upon his death in 1926, his son continued the search, and the third-generation heir, Mizuno Tomoyuki, dug around the mountains of Gunma until his death in 2010. Although much of this story became popular during the 1930s, legends of "military funds" being buried in neighboring Ibaraki Prefecture date back to the Meiji period. An article from 1880 follows a man who spent three months digging day and night in a local village searching for gold bars supposedly placed there by the Tokugawa family.[144]

Anthropologists have theorized that buried-treasure legends reflect economic conditions within a community undergoing transition. Within treasure narratives, a supernatural economy brings hope to the local community when the natural economy disappoints, promising "both quick wealth and a sense of power over the natural world."[145] Treasure legends also explain the sudden wealth experienced by one member of a community who becomes rich through nontraditional means, especially in villages undergoing the process of modernization.[146] Treasure hunting and fears of conspiracy, for example, arose during the French Revolution in places with rigid subsistence economies where oral transmission acted as the main vehicle for spreading knowledge.[147] One foreign resident who lived in Japan from 1870 to 1874 recalled stories of farmers who unearthed gold coins: rumors that reflected anxiety about farming in the modern economy. The observer had little sympathy, "So in Japan I met, while there, several foolish people whose whole mind was set on getting suddenly rich by finding buried money. The amount of spade-work and field-digging which they accomplished without any success would have sufficed to have made good farmers of them. It is a surer thing in Japan, as in America, to seek to find gold by steady work and a mind on the lookout for opportunities than by digging for it at random."[148]

These buried-treasure narratives have provided further dimension to the appropriation of Oguri's legacy. The story of the Tokugawa buried gold and its association with Oguri fits into broader, moral parables found in other treasure legends around the world. In his study of treasure legends, John Crossan divides such tales into two acts: the concealment of the treasure and the search.[149] During the concealment phase, the treasure is often buried in response to a threat of invasion.[150] Oguri, in variations of the legend, is ordered to bury money out of fear of the

West, as both a potential invader and the endpoint for Japan's hemor-rhaging gold bullion, or to protect against the shogunate's domestic en-emies, the Satsuma and Chōshū domains. Treasure buried underground usually indicates some degree of superstition in the story, especially the association between death and the earth. In the Oguri legend, laborers were killed and entombed with the gold after completing their work.

Treasure-hunting legends impart a moral lesson. The treasure is never found when actively sought because the stories usually include a dichot-omy of good versus evil, poor versus rich, virtue versus greed, or in this case the marginalized Oguri memory versus the pro-Satchō version of Meiji history. When money is found in the Oguri treasure legends, it is discovered by accident—for example, the local merchant who discovered money in the bottom of the miso jug. In one version of this story, the mer-chant visited Oguri's grave to keep from angering Oguri's spirit.[151] Another version claims that he died suddenly because he kept the gold for himself and suffered the anger of Oguri's ghost.[152] Even the West appeared on the butt end of the treasure legend. Oral history recorded in Akagi Village during the mid-1920s points to a group of laborers led by an American from Yokohama, who spent the spring and fall in 1873 digging holes in the mountain using a large machine never before seen by the local residents. Rumors spread that Nakajima sold the rights to dig in the mountain for fifty thousand yen, no small amount for the time, and that the American tried to sue once he learned he had been tricked. In this story the Ameri-can is in a position of weakness, duped by locals, and weighed down by a powerful machine unable to save him from financial loss.[153]

The legend also targets the early Meiji government and its legacy. Local historians and memory activists argue, quite logically, that the legend stems from the initial riot and Oguri's subsequent arrest.[154] Months after his execution, the monk of the Fumon'in Temple in Ōmiya City, Saitama, was beaten to death with a club. Oguri visited the temple on his way to Gonda Village from Edo and entrusted to the Fumon'in monk several heirlooms belonging to an ancestor who had renovated the temple. Some local people believe that the monk was killed because of his connection to Oguri. A decade later, in 1877, rumors spread that a local man's new-found wealth came from Oguri's treasure, acquired after torturing the Fumon'in monk.[155]

Conclusion

At the national level, only the community of former Tokugawa men had a personal stake in speaking about or valorizing Oguri, and when they died, as did Kurimoto, Fukuchi, Fukuzawa, and Katsu during the 1890s and early 1900s, interest in Oguri declined. Eventually the focus of the Meiji-era dialogue changed after the First Sino-Japanese War (1894–95). The younger generation of those like Tokutomi Sohō no longer focused their attention on a critique of the Satchō clique but instead began to support Japan's imperialist expansion of the 1890s onward.[156] During this time, more compelling figures for rehabilitation and appropriation arose, such as Saigō Takamori, Sakamoto Ryōma, and Yoshida Shōin. Oguri shared characteristics with these tragic heroes; both he and Sakamoto died violently before the Meiji period, thus freeing them from the stigma associated with Meiji politics. But other heroes were more suitable within the political trends of the Meiji period. For example, Sakamoto was young, adventurous, and, more importantly, not a high-ranking shogunate samurai. He and the young nineteenth-century activist Ōshio Heihachirō were appropriated by those in the people's rights movement because, as low status figures, they represented the "people."[157] Saigō Takamori became the most appropriated figure during the first half of the twentieth century by a broad range of groups. Pundits saw him as someone who fought against the ineptitude of the Meiji government, and before WWII, he was the one figure most frequently associated with cries for a second restoration.[158] More importantly, Oguri's legacy lacked material sites of memory that would have enhanced his presence on the memory landscape. In contrast, Saigō's popularity benefited from kabuki plays, artwork, and, most importantly, his statue in Tokyo's Ueno Park, which supported the textually mediated collective memory about him. The same could not be said about Oguri during the Meiji period.

Even without the physical commemoration to sustain interest in him at the national or local level, the foundational stories about Oguri affected his later praise. As with Saigō's legacy, people throughout the Meiji era projected their own interpretations onto Oguri: for Tsukagoshi, Oguri was a progressive; for Fukuzawa, he represented the traditional spirit.[159] Tsukagoshi and Fukuchi established the textually mediated collective memory that was copied by others and became part of the narrative and valuation of Oguri's life. Throughout the twentieth and twenty-first centuries, for example, those who wished to disparage Katsu or praise Oguri repeatedly invoked Fukuzawa's *On Fighting to the Bitter End*.

Oguri's presence in the historical memory of the Meiji Restoration tells us something about how the contentious national story was written. The strongest and most influential writing about Oguri came from village elite Tsukagoshi and from those who knew him well—the personal and the local—whose voices also found a place within the national narrative. Once Oguri's colleagues died, and critiques of the Satchō clique waned, it was up to local people to promote him not only as a national hero but as a regional one as well. Memory activists eventually accomplished this in the following Taishō period (1912–26) during the fiftieth anniversary of the Yokosuka Naval Base. The base was a potential site of memory about Oguri in the Meiji period, but it was only during the Taishō period that it became infused with such meaning.

Oguri's rehabilitation in the countryside was less salient than at the national level and occurred much later in the Meiji period. Like Tsuka-goshi, who studied under Oguri's local confidant, Satō, other local elites might have felt compelled to commemorate him. But Oguri spent little time in Gunma and then only in Gonda, where his legacy was clearly fraught with tension. After all, many villagers in and around Gonda suffered as a result of Oguri's presence, due to the riot against him and the sudden appearance of hundreds of local samurai sent to arrest him and his supporters. Moreover, the disorder experienced in the region immediately following the Restoration contributed to an atmosphere of fear and confusion that helped stimulate rumors and legends about Oguri. Only during the 1890s, when prefectures like Gunma and Nagano were contributing to local and national identity, did Gunma historians seize on Oguri to connect the local to the national. Either way, both the communicative and the textually mediated memories from the countryside later conditioned how Oguri's legacy spread beyond Tokyo and the Kantō plain.

CHAPTER 3

Redeeming Villains

The textually mediated memory established from 1868 to the 1890s defined how Oguri could be appropriated: as a paragon of bygone Tokugawa values, an internationalist, a progressive, and a tragic figure whose loss should be mourned. But other than a few isolated instances of commemorating him at the Tōzenji Temple in Gunma, there were no large-scale efforts to commit time, resources, and people to locate him, physically, on the memory landscape. This chapter addresses the early development of that landscape—how memory activists created objects and participated in ceremonies that anchored the past to the present, thus creating a collective memory about Oguri. But at a larger level, it also argues for the primacy of material and physical commemoration over written texts in promoting historical figures. Many of the new writings about Oguri, or Restoration losers more generally, occurred after commemorations were held, statues were built, and supporters fought to have their heroes recognized by awarding them posthumous court rank. Some of those new works, in turn, affected commemoration.

Material objects and commemorative activities reinforce collective memory and make it work; written words are not enough. The success of any memory project requires material elements that impact the senses and provoke emotions to support written discourse. Moreover, the effective use of these material elements is often a critical element in the competition for legitimacy of one historical interpretation over others. Fujitani demonstrates this in his utilization of Pierre Nora's concept of "sites of memory" to examine the museums, national holidays, and ceremonies that the Meiji state established to discipline popular memory and promote a history with an unbroken imperial line.[1] The Meiji state trained the senses and emotions, as Fujitani shows, to eventually create dominant, mainstream views of Restoration history favoring pro-Meiji heroes and narratives over Tokugawa ones. In so doing, however, the state provided the grammar and vocabulary of commemoration that others appropriated.

Supporters of the Tokugawa legacy, for example, used material commemoration to challenge the Meiji state's domination over history and memory in their celebration of Tokyo's tricentennial anniversary in 1889. Jason Karlin's study of the tricentennial anniversary demonstrates how the event created a sense of unity among all levels of Tokyo society by celebrating Edo period culture and history. He illustrates how former Tokugawa retainers preserved the everyday life and cultural memory of the Edo period to resist the Meiji state's "official narratives of the past."[2] The tricentennial celebration challenged state-centered commemorative activity and illustrated the multiplicity of actors contending in the production of history and memory.[3]

Throughout this chapter, I trace the work of memory activists through a broad range of objects and ceremonies. The account is neither top-down nor bottom-up. Local people often initiated statue building and commemoration; they used their proximity to the deceased to claim a sort of geographical authority to represent their hero. But to promote him successfully, they needed cooperation from sympathetic politicians, journalists, and scholars at the national level, all of whom found common ground in their support for Meiji Restoration "losers"—not just Oguri but also Ii Naosuke, the Aizu samurai, and others. Local activists who cooperated with others to acquire posthumous court rank (*zōi*) for Oguri best illustrate this interaction.

Perhaps most importantly, holding commemorative ceremonies and erecting statues produced a powerful, emotive discourse that expanded on the project begun with written words. The rehabilitation of the Restoration losers required a deconstruction of what made them "losers" in the first place, followed by the creation of an alternative narrative. This could offer opportunities for local groups to push ahead with their own agendas and force the government to play catch-up. The commemorations of Ii and Oguri at, respectively, the Yokohama and Yokosuka fiftieth anniversaries, for example, brought the two into the national spotlight and prompted memory activists to create sculptures to mark the occasions. In Ii's case, the controversy surrounding his statue not only provoked at least one supporter of his enemies to write a counter-narrative but helped prompt the launching of a committee, under government auspices, to publish biographies of those officially determined to be proper heroes. Morcover, it was only *after* memory activists tried to

erect a statue to Ii in a Tokyo park that the government created a law
limiting access to public commemorations in Tokyo.

Renewed interest in these figures continued to cause conflict over
subsequent decades as new interpretations of putative villains clashed
with the dominant, pro-Satchō view of Restoration history. Even among
supporters of the "villains," there was contestation over whose narrative would prevail. In Oguri's case, where lacunae in the historical record
encouraged invention, rival activists competed over who could legitimately speak about Oguri. Yet rivals attacked those who appropriated
him incorrectly—the treasure hunters who capitalized on the economic
desperation experienced during the 1930s and took advantage of the new
Oguri boom.

Commemorating Villains in Sculpture and Ceremony

Public space became contested terrain for those who wanted to commemorate heroes, but not all public space was created equal. From the
national point of view, public space was arranged hierarchically: a park
in Tokyo had more symbolic value than one in Yokohama, and Yokohama's national profile was higher than that of Yokosuka. Memory activists
attempted to take advantage of the new symbolism infusing public space
by erecting objects commemorating vilified figures whom they believed
to be significant in national history. The central government eventually
sought to reassert control over this effort by restricting individual and
localized memories in Tokyo, Kyoto, and Osaka, reserving those cities
for monuments deemed important in national and imperial history.[4] In
other places, however, material commemoration of contested heroes was
complicated by the dual nature of space that informed local identities in
addition to national ones. Here, timing mattered. When the government
celebrated Yokohama for its role in national history during the fiftieth
anniversary of the opening of Japan's ports, local commemoration of Ii
Naosuke became a threat. Once the celebrations were over, the city re-
verted to a local space, and the presence of Ii's statue no longer clashed
with the goals of the state.

That the state was an issue in the first place illustrates the significance
that statue building had assumed by the end of the Meiji era. Statue building began as a site of memory that valorized men who built the modern
nation-state, endowing their heroic legacy with a physical presence on
the memory landscape. Few such monuments had been erected early
in the Meiji period, and those that had been, such as the statues of Saigō
Takamori, Ōmura Masujirō, Yamato Takeru, and Kusunoki Masashige,

emphasized imperial loyalty and military prowess.[5] Saigō's bronze was potentially controversial because of his role in the 1877 rebellion, but the government hoped to dominate Saigō's overwhelming popularity by erecting a statue that deemphasized his role in the rebellion.[6] He was, after all, a colleague who had led the Meiji forces during the Restoration, and his statue faced Edo castle to reflect this; but the oligarchs supported only a pacified version of his legacy, lest he become a model for resistance.[7] Unlike the other pro-emperor statues, which display stern martial features, Saigō's 1898 statue in Ueno Park is singularly bucolic and, much to the chagrin of his widow, somewhat undignified.[8] He appears to be smiling, is wearing a casual robe and carrying a small dirk, and has an unimposing dog by his side. In his homeland of Kagoshima, a more imperious statue of Saigō dressed in Western military garb was erected in 1937 in Kagoshima's central park, asserting local views about his legacy in Restoration history. Saigō's image, in effect, switched places: appearing countrified in the national city while remaining a national figure in the countryside.

To the Meiji oligarchs, however, Ii Naosuke was an old enemy. He was perceived as having bowed to Western pressure while maintaining an uncompromising stance regarding the shogunate's preeminence as the sole governing body in Japan. Ii's vilification had been the most intense of all shogunal figures, even before the Meiji Restoration, among pro-imperial zealots, xenophobes (although Ii himself had no love of the West), and supporters of the opposing Hitotsubashi faction for shogunal succession. Purging political opponents in the late 1850s proved to be a fatal decision for Ii when renegade samurai assassinated him in 1860. As described in the previous chapter, Ii's role in the Tokugawa regime made him a rallying figure after the Meiji Restoration among Tokugawa supporters who hoped to save their own legacy by rehabilitating his image. However, Ii, like Oguri, was more than simply a national figure; as a former daimyo, he was a local one as well.

Beginning in 1881, Ii's supporters—including his future biographer, Shimada Saburō, the son of a Tokugawa vassal—planned to erect a memorial to him in Tokyo but were blocked by Home Minister Shinagawa Yajirō. Shinagawa pointed out that Ii had killed many imperial loyalists, such as Yoshida Shōin, Shinagawa's teacher and mentor. Shimada then tried to appeal to statesman and one-time Kanagawa governor Mutsu Munemitsu, claiming that Ii's continued vilification was rendered meaningless by the realities of modern Japan. After all, Shimada argued, men in power like Shinagawa, who once advocated "expelling the barbarians," now followed in Ii's footsteps and supported an open country.[9]

Mutsu, himself a former activist in overthrowing the shogunate, agreed to support Ii's commemoration, but even he failed to convince Shinagawa. Years later, in 1899, Toyohara Motoi, a former retainer of the Ii family, started a campaign to lease land in Hibiya Park to erect a statue of Ii. Again, Shimada joined the cause, as did historian Taguchi Ukichi, who often wrote about both Ii and Oguri. City hall officials agreed to lease the land, but perhaps hoping to avoid controversy, they pushed the statue issue to the Tokyo governor's office. In May 1900, after months of waiting with no response, Shimada and his group were informed of a new law that required anyone wishing to erect a statue on public grounds in Tokyo, Kyoto, or Osaka to obtain permission from the home minister; the new law effectively ended any chance that Ii would be commemorated in what the government defined as "national" cities.

It could have been the case that Meiji oligarchs were rejecting Shimada and Taguchi as much as they were denying Ii's new status as a hero. In addition to their longtime support for pro-Tokugawa commemoration, both men supported Ōkuma Shigenobu and, like Ōkuma, often criticized the Satchō clique. Taguchi, an economist, a naval man, and a Diet member, had been critical of the government's protective economic policies. His historical works, although not as widely read as those of his contemporaries, attacked the foundational Shinto myths that informed the political value of the imperial house.[10] Whatever the true reason, the central government would not compromise its hold over Tokyo public space as a site for controversial historical memory.

The Meiji oligarchs represented by Shinagawa Yajirō defeated Ii's supporters on the national memory landscape, but they could not directly control public space in Yokohama. In 1884, a former Hikone domain samurai from Ii's homeland purchased land in Yokohama and designated it as the place for Ii's statue and local commemoration.[11] Plans to erect the statue began in 1903 with donations from Japanese and foreigners living in Yokohama; in 1907, the governor and local police gave permission to raise the statue. The city intended to unveil it during the fiftieth anniversary of the opening of the ports on July 1, 1909.

The erection of Ii's statue on the anniversary at Yokohama, a port symbolizing Japan's modernity, pitted local interests against those of some of the Meiji leadership. The statue, needless to say, provoked controversy among those who supported contending narratives of the Meiji Restoration years. Some argued that Ii was not responsible for opening ports but had simply continued policies already begun by his predecessors Hotta Masayoshi and Abe Masahiro. More problematic were the elder statesmen (*genrō*), former Satsuma and Chōshū samurai, who con-

tinued to vilify Ii. They ordered Sufu Kōhei, governor of Kanagawa Prefecture, to stop the unveiling ceremony. Sufu, himself a former Chōshū
samurai who might have shared the anti-Ii sentiment, was shocked by
the demand but postponed the unveiling for ten days lest the invited
dignitaries refuse to attend the celebration.[12] In this way, opponents of the
unveiling temporarily succeeded in removing Ii as a potential symbol of
modern Japan.[13]

His statue also provoked a writing boom among those who wanted to
counter his new, heroic status. In 1910, a group formed within the Ministry of Education both to showcase the writings of and to publish biographies about Restoration imperial loyalists.[14] The strongest and most
immediate reaction came from Iwasaki Hideshige, who feared that by
heroizing Ii, the Mito rōnin would be vilified. To prevent this, an out-
raged Iwasaki published a heroic account of the Mito rōnin's actions,
which was purchased by the Ministry of Education and distributed to
elementary schools as a teaching material in 1912; thus, the government
had picked a side.[15] Critics complained that the government was promoting the assassination of a government minister as a praiseworthy
example of self-sacrifice. Iwasaki and others defended the government's
support, stating that because these Mito rōnin were enshrined at Yasukuni, the government had already acknowledged the rōnin. Yasukuni,
it seems, erased accusations of criminality then as it does in present-day
Japan. Moreover, they argued that the book's purpose was to defend
the Mito rōnin reputation against former Hikone men who wanted to
vilify them.[16]

Ōkuma Shigenobu was the lone statesman to give a speech at the unveiling ceremony, expressing his regret over pressure by the government
to postpone the unveiling. He criticized the previous emperor and *jōi*
advocates, especially those from the Chōshū domain, for their ignorance
of world affairs and refusal to open Japan. Ii, Ōkuma argued, had worked
for the good of the country as a whole, not just for the emperor, as did
Prime Minister Itō Hirobumi and others.[17] City officials felt that Ōkuma
had gone too far, and they removed his comments about the emperor
from subsequent publications regarding the 1909 commemorative events.

Ōkuma's contribution to Ii's and Oguri's commemoration dovetails
with his long-standing political clash with the Satchō oligarchy. Immediately after the Restoration, he butted heads with Saigō, refusing to appoint Saigō's favorites to positions within the Finance Ministry. He also
advocated changing Japan to a centralized state (*gunken seidō*), as Oguri
had once advised, a plan that Saigō opposed. Ōkuma's call for popular
participation in a constitutional government, modeled on the British

system, earned him the ire of more gradualist-minded oligarchs but made him the darling of many who felt he could stand up to the government and fight for the people.[18] Ōkuma's support for the existence of political parties and his critique of clique government put him at odds with other elder statesmen.[19] In his *Fifty Years of New Japan*, he described the Satchō clique's longevity after the establishment of the Imperial Diet in 1890 as the "co-existence of a clan executive and a national legislature [that] gave a peculiar aspect to the constitutional life of the empire and evoked many political embarrassments."[20] His various political stances led to his being ousted from government in 1881, and although he remained active in political circles with his Constitutional Reform Party, his newly created Waseda University, and his renegotiation of the so-called unequal treaties, he did not return to government until 1896. By the time he wrote his *Fifty Years of New Japan* and spoke at the Yokohama fiftieth anniversary, he had temporarily retired from politics and would engage in a series of cultural pursuits before returning to politics on the eve of WWI.

Ōkuma's professional troubles mirrored those suffered by Ii and Oguri. All three endured harsh criticism for their interaction with the West and hard-line approach toward domestic opponents. Oguri and Ii lost their lives; Ōkuma, a leg during a failed assassination attempt and, temporarily, a political career. Ōkuma's sympathy for Ii was clear; both men signed treaties with the Western powers; in fact, Ōkuma's was an unpopular revision of those signed by Ii. He lamented Ii's murder: "The price Japan had to pay for her progress and development was a dear one. . . . [I]t is well to remember that for the recent changes and its present position Japan had to make all of these sacrifices."[21] He also identified with Oguri due, in part, to their similar career paths. Ōkuma led both the Ministry of Foreign Affairs and the Ministry of Finance, positions once held by Oguri during the 1860s. He also paid off the shogunate's debt to the French government for its investment in the Yokosuka shipyard. Other writers at the time connected Oguri and Ōkuma more directly, such as Miyake Setsurei, who said, "Oguri was very much like Ōkuma. [A]t the time there was nobody who could compare to him, and nobody would listen to him; eventually he was killed by the emperor's army."[22]

Political motivations and a parallel professional life alone do not fully explain Ōkuma's effort to champion Oguri; in fact, his relationship to Oguri's family was direct and intimate. Ōkuma's second wife, Ayako, was Oguri Tadataka's niece from his birth family.[23] Ōkuma also took in Oguri's wife and daughter, Kuniko, after the death, in 1877, of their caretaker, Minomura Rizaemon.[24] He helped Kuniko find a husband, introducing her to Yano Sadao, the younger brother of Yano Fumio, editor of the *Yūbin hōchi*, a mouthpiece for Ōkuma's political views.[25] This was no trivial point for local memory activists in Gunma who appropriated Sadao, and no doubt his reputation, by having him submit articles on the Oguri family to local publications.

Yokosuka and the Oguri Boom: Commemoration in Deeds and Words

Oguri's legacy, its promotion at the fiftieth anniversary of the port city Yokosuka, and the erection of his bronze bust illustrate a different relationship between commemoration and the use of resources than the Ii Naosuke/Yokohama case. Although still a celebration of the nation, Yokosuka's anniversary was largely a naval and municipal affair. Ōkuma was the highest profile statesman in attendance, but he did not preside over the events, a responsibility left to Vice Admiral Kuroi Teijirō, director of the Yokosuka Naval Base and commander of the Yokosuka fleet. Moreover, no evidence suggests that Oguri was a controversial figure among high-ranking navy men, many of whom praised his contributions to the navy. Adachi Ritsuen's 1905 *Historical Discussion of Naval Charting (Kaikoku shidan)* includes a short, one-page description of Oguri's "great success in founding the navy which we must remember."[26] As with other Meiji-period Oguri commentators, Adachi reminded his readers that more popular Meiji figures, such as Katsu Kaishū and Ōkubo, were Oguri's subordinates. Adachi's assessment of Oguri focused largely on his relationship to Yokosuka, contending that it demonstrated how Oguri had challenged the accepted thinking of his time. Most naval publications avoided the controversy surrounding Oguri's support of the Tokugawa regime and his death, but not Adachi: "Although we should feel pity for Oguri's end, despite his association with the defeated government, we should honor him as someone who established the foundations of our country."[27]

Tōgō Heihachirō, hero of the Russo-Japanese War (1904–05), was the first government official to praise Oguri for having the foresight to plan what eventually became the Yokosuka Naval Base. Tōgō himself had served as the base captain in the late 1880s, early in his naval career, and the city responded to Tōgō's naval victories over the Russians by dedicating their victory celebrations to him. In 1912, Tōgō invited Oguri Sadao and Sadao's son Mataichi to his residence, where he expressed his opinion that Japan's victory during the Russo-Japanese War was largely due to Oguri's construction of Yokosuka.[28] To commemorate their visit, Tōgō wrote two sheets of calligraphy, which became markers of legitimacy for the two main contenders for Oguri's legacy during the 1930s and 1940s.[29]

Yokosuka naval and municipal historians redeemed Oguri by highlighting his contribution to Japan's military prowess, while his commemoration during the Yokosuka anniversary on September 27, 1915, brought him into the national limelight. Over two thousand people attended the day's ceremony. Newspaper articles covering the event highlighted Oguri as Yokosuka's primary benefactor, relegating the French architect Verny to a secondary role. In the September 28, 1915, edition of *Kokumin shinbun*, two separate articles covered the events, one a basic description of the celebration and another titled, "Fifty Years of an Open Port: Oguri Kōzukenosuke Built Japan's First Naval Yard, Now One of the Biggest Shipyards in the World." Naval publications in 1915 likewise rekindled interest in Oguri's image. The three-volume *History of the Yokosuka Naval Yard* (*Yokosuka kaigun senshōshi*), published by the Yokosuka Naval Yard, emphasized Oguri's efforts over others who worked to create Yokosuka. The eight-person Yokosuka shipyard committee was simply referred to as "Oguri and the committee" despite, for example, Kurimoto's vital role in introducing key French people to Oguri. A small pamphlet published in the same year included a short Oguri biography that drew on Fukuchi's book and Kurimoto's Yokosuka memoir, in which Oguri was portrayed as an official who did not believe in the impossible and whose resolve and financial abilities led to Yokosuka's completion.

Unlike the naval publications that simply praised Oguri for his contributions to the navy, works by Yokosuka municipal historians used the anniversary as an opportunity to rehabilitate Oguri's controversial past. A section titled "Origins of the Foundation of the Yokosuka Foundry" in Yokosuka City Hall's 1915 edition of *A Guide to Yokosuka* (*Yokosuka annaiki*) dealt mostly with Oguri's history. The authors denied the most typical accusation against Oguri—that he had been a Tokugawa absolutist: "He was troubled by the state of affairs and believed that the shogunate should be dissolved. Instead of wasting the country's finances, he wanted to build something that would last forever. He rejected arguments against him, used every day to serve the shogun, and decided to build the Yokosuka shipyard. We should remember that Oguri's insight is the reason why our country started its shipbuilding activities, and why Yokosuka exists as it does today."[30] Oguri, in other words, was a contributor to Japan's modernity, not simply a loyal Tokugawa retainer.

The anniversary presented memory activists from Gunma Prefecture with a unique opportunity to promote themselves and their own views on Oguri as a national hero, taking advantage of their localness as a source of legitimacy. They comprised the most active Oguri supporters at the anniversary, just as local people from Hikone had promoted Ii Naosuke in Yokohama five years earlier. At least ten people from Gonda Village attended, including village officials, a member of the board of education and Gunma Prefecture government, and the Tōzenji Temple monk. They brought with them objects related to Oguri to display during the celebration—resources that both supported the landscape of Gunma collective memory about Oguri and provided his, and their own, legacy with a physical presence in Yokosuka.[31]

In addition to monopolizing Oguri objects, Gunma historians could write Oguri's history in ways unavailable to professional historians— only natives, they could claim, had access to knowledge about Oguri's life and death in Gonda. Among the Gunma-born attendees was Tsuka-goshi Yoshitarō, who authored a pamphlet distributed throughout the crowd during the Yokosuka celebration. This pamphlet, *A Record of Oguri Kōzukenosuke's Last Days* (*Oguri Kōzukenosuke matsuro jiseki*), high-lighted two parts of Oguri's story ignored in most Yokosuka anniversary literature: Oguri's interaction with Gonda Village and his death. Since Oguri's putative criminality originated in his activities around Gonda, Tsukagoshi, as a native, enjoyed a more authentic role in rehabilitating Oguri's image than did outside historians, especially regarding Oguri's death. For example, Tsukagoshi hoped to dispel common misunder-standings about Oguri, such as his intention to launch a countercoup by first establishing a peasant militia. Tsukagoshi explained:

> The biggest source of misunderstanding about Oguri's sad end really has to do with his plans to establish a peasant militia. It was something incidental made inevitable. . . . [H]e was a man in the prime of his life, when he worked the hardest, so establishing a peasant militia after arriving in Gonda would have been part of his former job as the army commissioner. There just aren't enough documents to clarify whether he intended to use this group to fight the imperial army, or if he was simply preparing for the future, as was the case with Yokosuka—building a house with warehouse attached.[32]

Writing for a national audience during a celebratory moment for the nation, Tsukagoshi was careful not to offend the nationalism used to re-deem Oguri. He blamed local unrest, not imperial government malice, as the cause of Oguri's execution. Bandits attacked, Tsukagoshi tells us, during a lawless time, when rumors spread about Oguri's wealth hidden in munitions trunks and pickle barrels. Kanai, leader of the gang that wanted Oguri's wealth, "threatened the villagers" and gathered a force of seven thousand headed toward Gonda. "This was a sad event for Oguri . . . and it led directly to his persecution by the imperial army."[33] Tsukagoshi noted that discontent festering in neighboring villages was the only reason the imperial army eventually persecuted Oguri after the riot against him: "Following the riot, the local villages went and apologized to Oguri, suing for peace. Even though the situation seemed peaceful, reconciliation was incomplete. Village headmen and others from places with little connection to Oguri complained to the Tōsandō army."[34] Only then did the imperial army send troops to investigate the situation and determine that Oguri's military preparations were a threat. Even after executing Oguri, Tsukagoshi emphasized, the government expressed regret: *Ooto Ryōtarō ordered that twenty-five ryō be given to the Tōzenji Temple for Oguri's memorial services. His unjustified death became known to the government* (emphasis in the original)."[35]

The Yokosuka anniversary also inspired Gunma Prefecture historians to promote Oguri locally. Toyokuni Kakudō, a Buddhist monk and founder of the regional journal *Jōmō and the Jōmō People* (*Jōmō oyobi Jōmōjin—JOJ*), adopted Oguri as a figure to appropriate for regional identity: "We Jōshū [Gunma] people were delighted to hear of this [Yokosuka] event."[36] Appealing to a wider audience forced historians like Toyokuni to reconsider how to interpret Oguri's contentious local past. Unlike Tsukagoshi, Toyokuni avoided implicating locals in Oguri's death so as not to alienate anyone in the growing local collective memory. He downplayed the riot against Oguri that other scholars saw as a precursor to Oguri's arrest and transposed the narrative of greedy local gangsters attacking Oguri for his money onto the Tōsandō army leaders, Toyonaga Kan'ichirō and Hara Yasutarō. According to Toyokuni, they took advantage of the antagonistic feelings toward Oguri and, "hearing about the large amount of money that Oguri brought with him [to Gonda], they took it upon themselves to go after him in the hopes of gaining profit. They persecuted him for being a traitor without even investigating him."[37] To promote Oguri as a tragic Gunma hero, Toyokuni needed a local villain who was not *of* the local. He believed that the Satchō oligarchy shared responsibility for Oguri's death, but focusing blame on the Tōsandō leaders, non-natives who had caused havoc during and after the Meiji Restoration, evoked stronger emotions in Gunma.

Yokosuka City also claimed Oguri as a hero, and it was there that the movement to create a bust of Oguri began. Ozeki Susumu, a high-ranking Yokosuka naval official and Gunma Prefecture native, sent an open letter to the *JOJ* requesting donations from his home prefecture to build busts of Oguri and Verny in Yokosuka. The bust would acknowledge Oguri's "understanding that a navy was one of our country's most pressing needs."[38] The *JOJ* editor published Ozeki's letter and petition, imploring his readers, "as country-loving Gunma citizens, [to] send donations appropriate to your means directly to the Yokosuka City mayor."[39] In Ozeki's request, he noted that although society recently acknowledged Oguri's and Verny's contributions, "it is regretful that the number of people who remember this has declined," and for this reason Yokosuka citizens decided to erect busts on a hill in Suwa Park.[40] Oguri's grandson, Mataichi, and an unnamed French clergyman unveiled the busts on September 29, 1922, seven years after the Yokosuka anniversary.[41] The event was well attended and included such dignitaries as a member of the Tokugawa family, Ōkuma Shigenobu's older brother (Ōkuma had died earlier in the year), and high-ranking officials from the army and navy. The French ambassador, Paul Claudel, compared Oguri's rehabilitation to Saigō's: "Today all the old quarrels are forgotten and the imperial government wishes to remove these historical fissures . . . and honor the memory of the rebel Oguri just as it does the rebellious figure Saigō."[42]

Erecting Oguri's bust marked a turning point in his commemoration by invigorating local interest in him. Between the 1915 Yokosuka anniversary and the 1922 bust unveiling, Oguri-related articles in *JOJ* averaged only about one per year, sometimes none at all, but between 1922 and the late 1920s, they increased to about one article per month. An Oguri biography featured in 1917 in the journal *Kantō Youth* (*Kantō no shōnen*) was re-serialized in *JOJ* throughout 1922, perhaps because Gunma citizens needed to be made aware of their newly created Gunma hero. The biography, by local researcher Hayakawa Keison, was the most thorough Oguri biography since the first one written by Tsukagoshi in the 1890s. The editors noted that Hayakawa's biography surpassed others because it addressed Oguri's role in regional history and thus expanded the scope of the Oguri memory landscape.

Despite similarities between the commemorations of Ii and Oguri during their respective anniversary celebrations, Oguri's case fostered a unique development within his network of memory activists. Ōkuma's participation in both celebrations demonstrates how the Meiji Restoration, although decades old by 1909 and 1915, was not a distant memory among oligarchs who had a stake in how the Restoration was remembered. He attended these events because he sympathized with Ii and Oguri, recognized their efforts to contribute to Japan's development, and wanted to fight against what he believed was an unjust portrayal of these men by his political rivals. The major difference between the celebrations and subsequent sculptures was the breadth of the geographical response to Oguri's rehabilitation by memory activists. People from Gunma Prefecture cooperated with supporters from Yokosuka to erect Oguri's bust, while Gunma, Yokosuka, and Saitama Prefectures all cast Oguri as their own local hero. This eventually led to a competition between Gunma and Saitama activists, and attracted unwanted attention from treasure hunters from across Japan. Moreover, Ii's story was generally well known, but Oguri's legacy required a more complete reconstruction, which fostered greater memory work in the countryside.

Promoting the Dead

By taking advantage of Oguri's newfound fame, memory activists hoped to complete his vindication by pursuing a distinction sought by activists elsewhere—in particular Aizu descendants who supported latent Restoration heroes: posthumous court rank. Posthumous rank was premised on the legitimizing power of the emperor. The earliest and highest posthumous-rank awardees included those who defended, created, or epitomized the emperor-centered ideology. In making its awards, the Imperial Household Ministry accepted recommendations from families, historians, politicians, and government ministers to initiate and support promotions, while memory activists used those same people to obtain rank for their heroes.

Although an ancient practice, awarding posthumous court rank took on a new significance after the Meiji Restoration with the resurgence of the imperial institution.[43] It became yet another site of memory employed to enhance the importance of the emperor, in addition to the museums, symbols, and celebrations outlined in Fujitani's work. The post-Restoration promotion system differed completely from its pre-modern predecessor. Before the Meiji Restoration, posthumous rank was awarded to the ruling class, mostly warriors, but the modern system included all "historical figures," those important to the state and those deemed significant by local authorities.[44] The first promotions in the 1870s were given to men with deep connections to the imperial institution: figures from the distant past (Nitta Yoshida, Kusunoki Masashige), loyalist daimyo, or scholars who contributed to pro-emperor ideology (nativists such as Kada no Azumamaro, Kamo no Mabuchi, Motoori Norinaga, and Hirata Atsutane). They were followed by a mass promotion of imperial loyalists in 1890 (Sakamoto Ryōma, Yamagata Daini,

Fujita Yūkoku).[45] Not all loyalists were treated equally, however. The dominant figure in charge of conducting research for these promotions, Tanaka Mitsuaki, ensured that men from his own Tosa domain received preferential treatment. From 1885 to 1888, Tanaka conducted research for promotions announced in connection to the proclamation of the Meiji Constitution. He subsequently entered the Imperial Household Ministry, becoming its minister, where he continued to sign off on promotion candidates. During the Meiji period, Tanaka personally selected nearly 60 percent of all awardees, and of those, 70 percent were from Tosa.[46] Non-loyalists, however, also received promotions. Among the 2,171 persons listed in the *Lives of the Great People Given Posthumous Rank (Zōi shoken den)*, approximately 50 percent were identified as loyalist "men of high spirit," followed by 26 percent defined as "those of scholarly or artistic value to the public welfare," a group that included some former Tokugawa shogunate officials and women.[47]

Despite the apparent generosity in the bestowal of court ranks, the process remained an arena for continued contestation over the Tokugawa-period heritage. For example, one notable exception to the court ranks given to Confucian scholars was Ogyū Sōrai. Hayashi Razan and Itō Jinsai received rank early in the twentieth century and were promoted again during the enthronement ceremony for the Taishō emperor (1915), but Ogyū remained ignored. Several scholars commented on the lack of recognition for Ogyū's achievement, but future prime minister Inukai Tsuyoshi brought popular coverage to the issue. After a day of enthronement ceremonies, he complained to reporters that he did not understand the criterion for choosing rank awardees; to him it seemed arbitrary. He believed that trivial historical issues blocked Ogyū's promotion—for example, Ogyū neglecting to use the word "great" when referring to Japan, or suggesting that Japan was equal, not superior, to China, as Hirata Atsutane had argued.[48] The government history doyen Mikami Sanji attacked Inukai by saying, "Not only is Inukai's criticism without value, his ignorance makes an enemy of the very imperial benefice bestowed upon him. It's outrageous."[49]

Similar conflict arose over the ambiguous line separating those from the Meiji Restoration who deserved posthumous rank and those who did not. Kimura Kametarō spent years gathering documents to acquire rank for his grandfather Sagara Sōzō. In 1867 and 1868, Sagara led a vigilante group, later called the Sekihōtai, to win support from locals to fight against the shogunate's troops. Several minor Kyoto nobles led the group, which received authorization from Kyoto only after the fact, and Sagara and his men rushed toward the Kantō region claiming to be the vanguard for the Tōsandō army. Without permission from the Meiji leadership, however, he promised local peasants that the new government would cut their taxes in half. Having overstepped its authority, the Sekihōtai was labeled a "fake imperial army," Sagara was executed along with seven of his men by order of Iwakura Tomosada, the same leader of the Tōsandō army who ordered Oguri's execution and in the same expeditious fashion.[50] "It breaks my heart that many who were junior to my grandfather received posthumous rank but for no good reason he is still being denied," lamented Kimura.[51] His efforts paid off—Sagara received posthumous rank in 1928 and was enshrined in Yasukuni Shrine in 1929. Kimura's position at the Imperial Household Ministry probably helped him win his case.

Not surprisingly, Meiji Restoration "losers" also faced problems gaining posthumous rank and even legitimate commemoration for their dead. Aizu families faced the worst discrimination from the new government, which prohibited them from even burying their dead until February 1869. Until 1876, surviving warriors in the Aizu domain, who bore the label of "rebel" during and after the Meiji Restoration, were forbidden to memorialize their dead, even while families of Satchō soldiers received compensation for men killed in battle who were commemorated as having died for the emperor.[52] In 1914, Aizu residents formed a veterans' association to memorialize Aizu war dead by having them enshrined as patriots in Yasukuni Shrine. This would be no easy task, as Yasukuni was originally founded to enshrine the souls of those who died fighting on the "emperor's side" during the Boshin War. The Aizu group successfully enlisted help from the Society for Historical Narration (Shidankai) to support its campaign, but failed to convince the central government to admit Aizu dead into Yasukuni.[53]

The Shidankai, established in 1890 as a semi-official organization to promote the exchange of information regarding the Meiji Restoration, became a resource for memory activists. When its leadership changed in the early twentieth century, the group's agenda shifted to critiquing the clique government and publishing histories that balanced the dominant Satchō bias in accounts of the Meiji Restoration.[54] Its two major works consisted of short biographies about participants on both sides of the Restoration. For example, the 1907 monograph *Who's Who of War Martyrs* (*Senbō junan shishi jinmeiroku*) lists close to six thousand men who died in battle from fighting, suicide, illness, or "mysterious circumstances" between 1843 and 1890.[55] More than just an isolated text, however, *Who's Who* served to connect local memorial services within a memory landscape that recognized the sacrifices of those contained within the text.[56] Eventually, the government stopped funding the Shidankai and it became a private organization, but its influence in historical circles, sympathies with the "losers" of the Restoration, and cooperation with local interests made it the ideal ally for those who sought to rehabilitate their otherwise marginalized ancestors within national history.[57]

The Yokosuka anniversary quickened local desire to clear Oguri's name at the national level. Gunma activists called on the Shidankai for help in their effort to clear Oguri's name by having him awarded posthumous rank, and the organization cooperated as it had with Aizu activists. Indeed, Oguri, his son, and five executed retainers were among those that the Shidankai had identified as unrecognized Tokugawa supporters worthy of glorification. Oguri's *zōi* was spearheaded from three directions: naval men associated with Yokosuka, national politicians, and the Gunma governor. Navy Minister Yashiro Rokurō appealed to then prime minister Ōkuma to award *zōi* to "Oguri and eight others" for their work in Yokosuka, building the iron foundry that was the foundation of the naval base. He sent the request in July 1915 in hopes of announcing the new court rank during the Yokosuka fiftieth anniversary. No official response is detailed in the Diet records, but the request was turned down. It might seem odd that Ōkuma, given his close connections to the Oguri family and various occupational sympathies with Oguri, would not grant him posthumous court rank. In the summer of 1915, Ōkuma and his allies, including Yashiro, who wanted to curb Satchō influence within the navy, were simply not in a position to bother with such a request. It had been a busy year for Ōkuma, starting with the infamous Twenty-One Demands sent to China, which had resulted in two treaties in May. The resulting international pressure paled in comparison with Ōkuma's domestic woes: his home minister, Ōura Kanetake, was forced to resign in July 1915 after he was caught trying to buy influence among Lower House members for Ōkuma's military spending bill. Ōkuma's cabinet resigned in the fall of 1916. Still, another attempt was made in May 1916, when Okada Keisuke, then head of naval personnel and a future prime minister, also sent a request to the cabinet secretary requesting rank for Oguri and eight others, including Kurimoto Joun. Then, in 1917, Shimada Saburō and Honda Susumu, a former Tokugawa vassal, sent a petition to the Shidankai enlisting its help in requesting that Oguri's name be cleared and that he be awarded posthumous rank.[58] Referred to as Oguri's "colleague," Honda had helped command the Shōgitai during the Meiji Restoration, and had been involved with historical memory through the journal *The Former Shogunate*.[59] Shimada used his newly appointed authority as Speaker of the House to submit the request.[60] The Shidankai and the Gunma Prefecture governor petitioned the prime minister's cabinet, while Toyokuni also entreated his local audience to support the movement: "If he (Oguri) were someone from Satsuma or Chōshū, especially an official, the government would have pulled out all the stops. Of course such a person would be awarded rank in addition to receiving other recognitions. We Gunma people should give our thanks to Kurui, head of the Yokosuka Naval Base, and never forget men like Shimada and Honda."[61]

As with the Aizu people's demands for recognition, however, the movement failed. Oguri Sadao, perhaps in a moment of sour grapes, put on the best face he could: "We should be thankful to those who tried to acquire posthumous rank for Oguri. . . . However, a person's value is not determined by rank but by the virtue given to them by heaven. Since the days of old, persons of value were honored as gods. . . . It followed naturally that a person of reputation would be known throughout society. [Simply] signifying someone as having such-and-such rank can instead be seen as an insult."[62] Sadao tried to claim the moral high ground for Oguri by subordinating the lesser, worldly court rank to the greater rewards from "heaven" and the "gods," the true arbiters of an individual's value. Subsequent publications by Ninagawa Arata, Toyokuni, and others echoed Sadao's sentiments.

While *zōi* efforts centered on the fiftieth anniversary at Yokosuka failed to produce results, another attempt was made during another significant year in Oguri's commemoration—1928. This was the sixtieth anniversary of the Meiji Restoration, a year when many pro-Tokugawa works were written, including the new Oguri biography described in the following section. This version of Oguri's life was widely read, prompting another campaign for his recognition. Unlike the 1915 campaign, this time there seemed to be less involvement from national politicians. Men like Shimada Saburō or Ōkuma, who had deep connections to the Oguri or Tokugawa legacies, had died. As before, the navy still appropriated Oguri as part of its legacy, and local people continued to be at the forefront of this effort. In 1928, Ichikawa Motokichi, the mayor of Kurata Village (formerly Gonda Village), sent a petition to the Gunma governor pushing for greater efforts regarding the issue, citing Oguri's contribution to the country, the navy's recognition of his achievements during the Yokosuka anniversary, and the subsequent creation of the Oguri bust.

A memo from the home minister to Prime Minister Tanaka Giichi details the Gunma governor's request for eleven promotions, citing the nominees' loyalty to the imperial country (*koshitsu kokka*). Oguri's rank was actually the lowest among those listed, the minor fifth rank (*jugoi shita*), which was typical for daimyo and prestigious bannermen.[63] The governor outlined a description of Oguri's accomplishments: his contribution to the navy, acknowledged by his bust in Yokosuka; his financial expertise; and his diplomatic abilities toward "Caucasian countries," referring to his time in the United States. Regarding Oguri's death, the author notes that other daimyo and bannermen, even Fukuzawa Yukichi, once advocated fighting against Satsuma and Chōshū; it was not only Oguri's opinion.[64] As for his being labeled a rebel, the donation by the Taishō empress to the instillation of his bust in Yokosuka proved that all was forgiven.[65]

In the same year, 1928, Navy Minister Okada Keisuke also wrote to Tanaka Giichi about Oguri, not to ask for posthumous rank but to explore the possibility of granting Oguri a special pardon.[66] This long document follows many of the same themes as the rank petitions, highlighting, for example, Oguri's contribution to the navy, and includes lengthy excerpts from works by familiar Oguri biographers, such as Fukuchi Gen'ichirō, and even from Seta Tōyō's youth reader. The government turned down Okada's request because it believed that Oguri's crime, namely that he suggested to the shogun that they fight against the imperial army, was pardoned under the general amnesty given to all former rebels, including Saigō Takamori, that had been announced with the proclamation of the Meiji Constitution.[67]

In each of these cases, Oguri's supporters understood the difficulties he faced due to his label as a rebel. They shifted the focus away from his actions in the countryside, noting that he was wrongly killed by those who misunderstood his intentions, and, instead, tried to normalize Oguri's past in light of how other Tokugawa retainers behaved during the Restoration; in other words, why single out Oguri for condemnation when other former retainers and daimyo were employed and even celebrated after the Restoration? This is how Okada understood Oguri's legacy. He did not seek to acquire rank for Oguri but wanted to erase undue focus on Oguri's putative crimes, a long-standing discrimination that separated him from those pardoned in 1889. He contended that being denied posthumous rank did not necessarily imply criminality, since many would-be candidates, including pro-emperor samurai, were also rejected. In 1944, the speaker of the Gunma Prefecture Assembly led the last movement to obtain Oguri's court promotion, but it too failed.[68]

Ii Naosuke and Matsudaira Katamori, former lord and a representative figure for the Aizu "rebellious" samurai, both received posthumous court rank. One might ask why Matsudaira had been promoted after his death in 1893 while Oguri never was. In the Meiji bias discourse, local historians who wrote about their domain's involvement in the Restoration tried to demonstrate how their predecessors ultimately fought in the name of the emperor. Matsudaira, who held the title of "protector of Kyoto," fit into the framework of loyal imperial retainer. Moreover, whatever resentment toward Katamori existed among the Satchō clique government, he lived well into the Meiji period and served as chief priest of the Nikkō Tōshōgu Shrine, complicating any attempt to continue vilifying him. In 1928, one of Katamori's granddaughters married Prince Chichibu, Hirohito's younger brother, in what was celebrated among Aizu residents as a long-awaited restoration of honor.[69] Ii Naosuke received posthumous rank but not until 1917, long after Katamori. Observers noted that although Ii was still considered an enemy among the Meiji oligarchs, his contributions to Japan were sufficient to warrant official recognition.[70]

The *Bakumatsu* Boom and Ninagawa Arata

Interpretations of the Meiji Restoration changed in the early Taishō period (1912–26), as domestic and world events drastically affected the intellectual environment in Japan: WWI, the Bolshevik Revolution, the 1918 rice riots, the 1923 Kantō earthquake, and the assassination of the prime minister. Ozaki Yukio, the father of parliamentary government, compared the Kantō earthquake to the 1855 Ansei earthquake and Prime Minister Hara Kei's assassination to that of Ii Naosuke; others dubbed the Russian Revolution shocks rippling through Japan the new "black ships" like the American ships that once imposed upon Japan. By the 1920s, the Meiji-period value system had faltered, and intellectuals felt both unease and hope for a new revolution to occur in Japan.[71] Government criticism continued alongside the spread of imperial democracy and mass culture, which shaped portrayals of Restoration figures in the Taishō period. Shirayanagi Shūko wrote a short biography of Sakamoto Ryōma, describing him as a forerunner in destroying feudalism who created a new Japan with the emperor at its core.[72] Yoshida Shōin changed from being a spiritual revolutionary in 1893 to a supporter of imperial expansion in 1908 and, by the 1930s, in Nakazato Kaizan's biography of him, a religious figure who wanted to save the masses.[73] The late 1920s also experienced a "Saigō boom," beginning in 1927—the fiftieth anniversary of his death. He too underwent various reconfigurations, as one among the many faceless citizens who rose to destroy past evils or otherwise fought for "the people."[74]

The Meiji Restoration's sixtieth anniversary, held in 1928, created another boom in popular depictions of the Restoration, exacerbated by a concern after the Kantō earthquake to preserve documents, thus encouraging the writing of new secondary works—in particular, those that praised the losers.[75] A series of interviews with a reluctant Tokugawa Yoshinobu finally appeared in 1925 after several decades of work, followed by a multivolume biography published in 1933.[76] Shimozawa Kan rehabilitated the Shinsengumi in three books published from 1928 to 1931, reflecting dissatisfaction with the clique government's unfulfilled promises set forth in the Charter Oath and the dominant historical narrative.[77] He rejected the typical portrayal of the Shinsengumi as a group of marauding assassins, represented in Nishimura Kenbun's 1894 account, and emphasized its role as a public police force, working for the collective and greater good of the nation.[78] Historical memory in Aizu also surged during the late 1920s and 1930s, perhaps due to their newfound rehabilitation. Yamakawa Kenjirō—a former member of Aizu's White Tiger Brigade, who served as president at several universities, including Tokyo University—was probably the most famous of Aizu's historical rehabilitators. His landmark *Aizu and the Boshin War* (*Aizu Boshin senshi*) was published in 1933, shortly after his death.

By the time this *bakumatsu* boom occurred, Oguri's popularity had reached an all-time high following the Yokosuka-inspired commemorations. Attention to Oguri continued to surge after a prolific law scholar and distant Oguri relative, Ninagawa Arata, became involved in Oguri's commemoration.[79] He entered into a mutually beneficial relationship with memory activists, obtaining information from native informants while providing locals with the means to bring their message to a national audience. Competing local activists sought to appropriate Ninagawa to legitimate their own claims over Oguri's legacy. In the process, Ninagawa's vitriolic criticism of Meiji heroes such as Saigō Takamori and Katsu Kaishū influenced subsequent Oguri depictions in the memory landscape.

Although the majority of Ninagawa's writings throughout the prewar period centered on his interest in diplomacy and Japan in East Asia, he published a two-part biography about Oguri on the sixtieth anniversary of the Restoration: the 1928 *Political Strife of the Restoration Years and the Death of Oguri Kōzukenosuke* (*Ishin zengo no seisō to Oguri Kōzuke no shi*), followed by a second volume in 1931. He did not present much new information on Oguri, but more than any previous biographer, he tied Oguri into a broader interpretation of the Meiji Restoration, one so laden with passionate hatred for the pro-Satchō narrative of Meiji Restoration history that it forever changed the textually mediated historical memory about Oguri. He even had trouble finding a publisher, finally convincing Nihon Shōin to publish the book by forfeiting his royalties.[80] He showed the pre-published manuscript to then prime minister and friend Tanaka Giichi, who told Ninagawa, "Shouldn't you stop sifting through the past with a toothpick?"[81] The publisher had legitimate reasons to be worried about the content of Ninagawa's book. The Peace Preservation Law of 1925 criminalized discussion of the national essence or capitalism, a measure aimed at Communist writings, a situation that only worsened during the 1930s.[82] Luckily for the publisher, controversy sold well. It became a best seller, going into a third printing after only ten days on the market, and was eventually published in sixteen editions.[83]

Ninagawa's reinterpretation of Oguri fell into two trends of the times. The first was the historical-memory trend: a resurgence in "loser history." In previous decades, many connected Oguri's legacy to the Aizu fallen, the Shinsengumi—particularly its captain, Kondō Isami—and Ii Naosuke. But Ninagawa directly cites new works written during the sixtieth anniversary Restoration boom and weaves them into the memory landscape that had expanded after the Yokosuka anniversary and the erection of Oguri's bust. His engagement with new intellectual trends is evident in Yamakawa Kenjirō's foreword to Ninagawa's book, which, as forewords are meant to do, reveals much about the intellectual agendas and alliances of like-minded authors. Yamakawa's commentary traces the deepening connection between Oguri and Aizu history when he explains that he recalled hearing about Oguri's mother and wife and their time in Aizu from the Yokoyama family, whose patriarch gave them refuge. Citing Oguri's contributions to the country, namely Yokosuka and the effort to create a centralized country, Yamakawa endorses not only Ninagawa's interpretations of Oguri but his revisionist view of the Restoration as well. Ninagawa returned the favor by drawing heavily from pro-Aizu works published during the sixtieth anniversary.

The second trend involved bringing Oguri's story up to date in the historiography by emphasizing his service to the state (*kokka*). Ninagawa employed a broad range of "statist" vocabulary at a historical moment when, to borrow Jennifer Robertson's words, *kokumin* was an omnipresent prefix.[84] Words like *kokumin* (citizens), *kokumin kokka* (citizen state), *kokumin bunka* (state/citizen's culture), and *minzoku* (ethnic group), which all have an intellectual history dating back to the late nineteenth century, reached their highest resonance during the 1930s and 1940s. They were not neutral terms that referred to Japan as a country but to an idealized, unified, ethnically privileged nation-state ready to take on all challengers. Ninagawa qualifies nearly every description of Oguri's career with these terms. Thus, Oguri "promoted state culture" (*kokka kokumin no bunka*), "wanted to foster happiness for the state and its people" (*kokka to minzoku*), and had always had "the peace of Japanese citizens" (*Nihon kokka kokumin*) in mind.[85]

In addition to his goal of demonstrating Oguri's value to the *kokumin*, Ninagawa wanted to "reveal the truth of Meiji Restoration history."[86] Misunderstandings about Oguri, Ninagawa argued, stemmed from a fundamental misinterpretation in the Satchō narrative. "Once Satchō unified the country, clique leaders did as they pleased and portrayed everything done by the Tokugawa as bad. This has been the view for the last sixty years. All elementary and middle school textbooks tell history from the Satchō perspective."[87] For Ninagawa, the Meiji Restoration as Japanese had understood it for the last sixty years was a lie.

He targeted famous Meiji heroes, especially Saigō. Both Saigō and Oguri were killed as rebels by government troops, Ninagawa noted, but the similarities ended there. "Oguri studied economics, finance, military affairs, and diplomacy and never bothered to write poetry. Saigō studied Wang Yangming Confucianism and even placed faith in spirits. He often wrote poetry, but cannot be considered a scholar. . . . Oguri was the foremost advocate of opening the country, he followed the trends of the world and wanted to create a centralized state. All we hear about Saigō are his opinions about the world around him."[88] Here Ninagawa points out how Saigō studied outdated Confucianism and was irrationally religious, while casting Oguri as the epitome of modernity, a master of economics, politics, and military affairs. For Ninagawa, Saigō did not even exemplify *bushidō* (the way of the warrior), because the arson and robbery that spread throughout Edo and the Kantō countryside at the hands of Satsuma men, and on Saigō's order, was the work of thieves and not *bushidō*-like.[89] Ninagawa held the Satchō oligarchy responsible for Oguri's death: "What do the Japanese citizens today think about Oguri's unjust murder by the imperial army . . . was that not the evil of thugs? Was this not against the Charter Oath's claim to follow the just laws of nature?"[90]

Commentators on both sides of the political aisle—such as Kōtoku Shūsui on the left, and Kita Ikki, the so-called "father of Japanese fascism," on the proverbial right—revered Saigō. Ninagawa fit into the latter group. "Unlike today's government, which is *based on public opinion* (emphasis in the original), Oguri followed autocracy, the virtuous government of his time."[91] Although he did not advocate a return to autocracy, Ninagawa explained throughout the book that it was the Tokugawa regime's understanding of the national policy (*kokuze*) that had allowed it to succeed for so long, and it would have been capable of modernizing the country. In Ninagawa's portrayal, Oguri lacked any political agenda, making him an ideal patriot in comparison to those who were deeply involved in politics. Unlike his lesser contemporaries, Oguri was above politics. During one tirade against Katsu Kaishū, in which he drew on Fukuzawa's *On Fighting to the Bitter End*, Ninagawa states, "Someone like Katsu, a master of flip-flopping even from the point of view of this age of party politics . . . he was dangerous and not to be trusted."[92] Ninagawa had no patience at all for leftist views: "Today the pacifists and socialists all curse military instillations. But the base [Yokosuka] saved 60 million Japanese people from the threat of the Chinese and Russians. It also helped bring peace to Asia and saved the citizenry (*kokumin*)."[93]

In the prologue to his second volume, Ninagawa addressed the controversies that arose from his first book. Many complained that Ninagawa misrepresented Restoration events, attacked Katsu and Saigō too much, and overly praised Oguri.[94] Undeterred, he reasserted his positions on both topics. Moreover, he included favorable book reviews and excerpts from letters received about his first volume, many of them written by scholars and other luminaries. Sano Asaō claimed that he had always called Saigō a "hero" but that after reading Ninagawa's book, he was not so sure.[95] Ōkuma Shigenobu's oldest daughter wrote to express her joy that Oguri's hidden contributions had been publicized.[96] Novelist Jūbishi Yoshihiko, inspired by Ninagawa's book, published a play in 1929 titled *The Death of Oguri Kōzukenosuke*.[97] Nakazato Kaizan had written about Oguri in previous editions of *Great Bodhisattva Pass*, but the 1928 volume, titled *Ocean*, directly addressed Oguri's new commemoration: "Oguri's is the one name that shouldn't be forgotten, but it has been forgotten for too long. . . . [I]t's not just that Katsu and Saigō suddenly outshined him . . . it's also that history is always written by the winners."[98] So too did Itō Chiyū credit Ninagawa's book for helping him understand the significance of Oguri, whom he included in his own 1931 pro-Tokugawa history *Great Supporters of the Shogunate* (*Sabakuha no ketsujin*).[99] Generally Itō was sympathetic, though he did point out that while some might claim otherwise, Oguri was not solely responsible for Yokosuka—Kurimoto did much of the work on the ground.[100] Science historian Mikami Yoshio wrote to the *JOJ* editor about Ninagawa's book, expressing anger over Satsuma deceit and joy over the government's crushing of the Satsuma rebels (referring to the Satsuma Rebellion).[101] The *JOJ* editors hoped that Ninagawa's book would rectify the public's misconceptions of the Meiji Restoration and restore honor to Oguri and Aizu domain men, demonstrating that even in Gunma Prefecture, Oguri's legacy was seen as operating within a larger memory landscape of Restoration losers.[102]

Not all historians agreed with Ninagawa's attempt to praise Oguri at the expense of received Meiji heroes. Kanakura Masami's 1935 *French Diplomat Roches and Oguri Kōzukenosuke* (*Furansu kōshi Rosesu to Oguri Kōzukenosuke*) questioned the contemporary accolades showered on Oguri for building Yokosuka. He agreed that Oguri's reputation should be positively reevaluated in light of Yokosuka's importance to Japan, but for him, the accolades were too generous because the foundry, arsenal, and dry dock failed to help the shogunate. "As a financial commissioner, Oguri just wasn't successful."[103] Kanakura appropriated Oguri not as a model for imitation in contemporary Japan but as an example of government wrongheadedness. He compared the growing military spending during the 1930s to that of the Tokugawa shogunate under Oguri: "We all want to improve the military, and build more planes and ships, but there has to be a balance between military power and the power of the people and the country. Without balance, one cannot take advantage of military power. It was the same with the shogunate building Yokosuka, it did not need it. Oguri was convinced by Roches' words. We can't really blame the shogunate though."[104] Kanakura also disagreed with the view that the imperial army killed Oguri without cause. "Some claim that Oguri was killed without having committed a crime, but this is a mistake. The imperial army punished anyone who did not clearly support the court. . . . [I]t also killed anyone causing chaos, even Satsuma's own Sagara Sōzō."[105] Kanakura believed that the imperial army had simply made a mistake; rumors about Oguri's putative anti-emperor stance spread throughout the countryside, and the army believed it.

The *bakumatsu* boom coincided with a growing interest in local history and folk studies. As Kären Wigen has recently demonstrated, the economic crash of 1929 hastened an effort to revitalize the countryside.[106] Local and long-standing sources of income were particularly hard hit in places like Nagano, the site of Wigen's case study, where agriculture and sericulture comprised a significant part of the local economy, a situation similar to that in Gunma. These changes brought about political agrarianism that at times called for another Restoration from the bottom up, from the provinces to the capital, to rectify the incomplete Meiji Restoration.[107] Another growing trend during the 1930s was efforts by scholars to promote a unique Japanese ethnic identity, one that would foster patriotism. Scholars like Yanagita Kunio looked to the geographical margins to find the purest manifestations of Japanese identity, putatively unblemished by urban modernity, one that was often elite and male.[108]

The trends that fostered native-place studies in Nagano had parallel developments in Gunma, which affected Oguri's legacy there. In Nagano, native-place history and geography became a pedagogical tool, acting as a mental bridge for students, first by connecting them to their native region and then by tying their region to the nation. Biographies of famous locals who made a name for themselves on the national stage, especially in military and educational affairs, were central to this pedagogy.[109] The efforts of Gunma educators paralleled developments in Nagano, and Oguri's legacy in Gunma was transformed accordingly. As just one example, the *Native-Place Reader* (*kyōdo dokuhon*), published in 1929 and republished in 1941, opened with a statement regarding its goal to foster "national spirit" (*kokumin seishin*) among schoolchildren through history and geography.[110] Oguri appears in his own chapter, which highlights his military accomplishments—namely, building Yokosuka and importing Western military techniques. The narrative does not explicitly blame the Satchō army for killing Oguri but suggests that his end resulted from animosity against him within the shogunate as well as misunderstandings of his intentions in the countryside, caused by gangsters looking for money. In any event, it ends on a positive note: the Taishō empress's donation for his bust in Yokosuka.[111] From the late 1920s into the 1940s, Gunma historical works increasingly appropriated Oguri to connect the local to the national, and national figures participated in local Oguri commemorations with greater frequency.

Memory activists turned to Ninagawa to help shape local collective memory. His rhetorical strategies influenced local discourse about Oguri, and Ninagawa himself participated in the constantly shifting local identity. Gonda Village no longer existed as an independent administrative unit. In 1889, it had joined with neighboring Sannokura Village to become Kurata Village, while Kawaura, Iwakōri, and Mizunuma—villages closely tied to Gonda as a corporate unit during the Tokugawa period—formed Ubuchi Village. Kurata and Ubuchi had cooperated on numerous educational and administrative projects since the nineteenth century. In 1931, elites from both villages formed an organization to create an Oguri memorial stone and named Ichikawa Motokichi, Kurata's mayor, as its president.[112] The elites asked Ninagawa for suggestions regarding what to write on the monument. He sent them two epitaphs; the first read simply, "The final resting place of Oguri Kōzukenosuke, a great man of the *bakumatsu* period." The second inscription, eventually chosen by the villagers, read, "Here lies the great Oguri Kōzukenosuke, killed without having committed a crime."[113] By choosing the latter inscription, village elites highlighted an aggressive interpretation of the Meiji Restoration during a time of growing ultra-nationalist ideology and violence. Authorities in Takasaki refused to allow the villages to erect the stone due to its controversial epitaph, which suggested unjustifiable homicide by the emperor's forces. The police argued that the imperial army would not have killed an innocent man.[114] Ninagawa apparently interceded on the villagers' behalf, and the stone was successfully unveiled a year later at a site near Oguri's execution ground. Over one thousand people turned up for the event, including officials from the Gunma Prefecture office, Ninagawa, Sadao's nephew (Sadao was ill), a descendant of Oguri's wife's family, and even the Takasaki police chief.[115] Ninagawa said that this was the first time villagers had expressed their gratitude to Oguri publicly: his spirit protected the region, "not as a government god, nor a god for self gain, but a god of the people."[116]

Getting a Head: The Local Fight over Oguri's Legacy

Oguri commemoration often fostered cooperation at the local level; many researchers collaborated, or cited one another's works, particularly in the *JOJ*, where much of Oguri's regional legacy was constructed. In the late 1920s, the Gunma Prefecture government obtained private donations to fund a relief of Oguri's bust to be placed in the Gunma Kaikan, an assembly hall located in the capital. The bust was completed in 1930, solidifying Oguri's importance to Gunma identity. Amateur historians throughout the prefecture studied Oguri's connection to their own areas, adding their history to the larger, regional one.[117]

But conflict, competition, and tension undermined local cooperation regarding Oguri's legacy. Commemorative activity revolving around the Tōzenji Temple in Gunma competed with that of the Fumon'in Temple in Saitama, both wanting to dominate Oguri-related memory. Even though their structures of collective memory clashed, drawing on competing resources such as access to local and national notables for support, they agreed on *how* Oguri should be remembered. Both sides attacked those who sought Oguri's gold, seeing it as an inappropriate use of Oguri's memory. These broad appropriations of Oguri's legacy arose from a lack of detailed information about his personal life and role in the countryside.

Competition among residents of different prefectures over who could legitimately represent Oguri in the countryside arose from growing

Restoration-related publications and commemorative activity during the
1920s and 1930s. As natural centers for memorial services, the temples
Tōzenji and Fumon'in fought over the right to appropriate Oguri. The
Fumon'in had an older connection to the Oguri family through the first
Oguri patriarch, who reconstructed what had been a dilapidated temple
in the sixteenth century. He and the first four Oguri patriarchs were buried in the Fumon'in, but for some unknown reason, the main family no
longer used the Fumon'in, having moved its family temple to Edo. When
Oguri Tadamasa left Edo in 1868, he stopped briefly at the Fumon'in,
donated 50 *ryō* for its future upkeep, and turned over several family heirlooms. Although the Tōzenji's connection to the Oguri family was not as
old, dating back to only the eighteenth century, it maintained a stronger
link to Tadamasa.

Fumon'in did not play a prominent role on the memory landscape
relating to Oguri until the 1930s, when Abe Dōzan pushed it into the
spotlight of Oguri commemoration.[118] According to Abe, who took over
the temple in 1925, people started visiting the Fumon'in after Tokutomi
Sohō published an article in 1933 about his visit to the "Oguri family
grave" located there.[119] In 1935, Abe unveiled a commemorative monument that attracted numerous celebrities, including Prime Minister
Okada Keisuke, Tokonami Takejirō, the Saitama Prefecture governor,
high-ranking naval officers, chief of the Yokosuka arsenal, and Ninagawa
Arata, who had initiated the project and written the words for the monument.[120] Okada's appearance was not surprising; after all, as navy minister
he had tried to obtain a special pardon for Oguri. Tokugawa Iesato, the
sixteenth-generation head of the Tokugawa family, wrote the calligraphy,
lending even more legitimacy to the Fumon'in while further defining
Oguri as a Tokugawa martyr. One newspaper article described the event,
in particular Prime Minister Okada's participation, as a rewriting of
Meiji Restoration history, a revisionist trend that began in the 1920s. The
anonymous author notes that among vilified Restoration figures, Oguri's
vindication had waited the longest. Itō Chiyū agreed: "Now that Oguri's
image has been rectified, there is nothing left to rewrite about Meiji
Restoration history."[121] His comment underscores Oguri's status as
the predominant Restoration loser, whose history marked the extreme
boundary for Restoration memory. Even Nakazato Kaizan visited the
Fumon'in Temple in 1936. According to his own account, he was driving
back to Tokyo through Ōmiya when he saw the sign "Fumon'in—The
grave of Oguri Kōzukenosuke's Head." He visited Abe and later wrote
about the temple's deep connection to the Oguri family and, of course,
Tadamasa himself.[122]

Fig. 2. Oguri's headstone at Fumon'in. According to the monks there, writing was purposely left off the stone in order to protect the grave from vandalism.

Oguri supporters in Gunma grew increasingly jealous of Abe's success in expanding his role within the Oguri memory landscape in Saitama. In 1940, he formed the Association to Promote Oguri Kōzukenosuke; the Saitama governor acted as president, members of the navy and Yokosuka Arsenal served as officers, and Ninagawa was listed as an advisor.[123] Throughout the 1940s, the group conducted an annual Oguri festival, widely attended by local citizens and people from Yokosuka. They also sent representatives to the annual Oguri-Verny memorial services held in Yokosuka. Gunma memory activists lamented Fumon'in's new fame: "Fumon'in is nothing more than the cemetery for Oguri's ancestors. The group there is a result of Abe's zeal. But Gonda in our prefecture (Gunma) is where Oguri['s] and Mataichi's remains are buried, even his retainers' graves are located here. We have Oguri Sadao's support, and the widow of Sadao's son and heir, Mataichi, has come to pay her respects at the graves many times."[124]

Tōzenji's only advantage over Fumon'in was its deeper connection to Oguri Sadao and his family, as Toyokuni pointed out. He suggested that Gunma was just as active, if not more so, than Saitama; Gonda too had

its own memorial association to help shepherd visits from various officials. In a rebuttal, Abe directly attacked the Gonda villagers for trying
to obtain recognition from the Oguri family and others important in
the creation of Oguri discourse. "The ungratefulness of the Gonda villagers struck at the heart of the Oguri family. Oguri's daughter Kuniko
said she would never visit Tōzenji."[125] He claimed that Oguri's wife,
Michiko, hated the Gonda villagers and accused them of betraying Oguri
by turning him over to the Takasaki domain. For this reason, Abe said
he would never visit Gonda.[126] Oguri Sadao's visit to the Fumon'in
graves boosted its legitimacy; Abe even received a letter from Sadao
stating that Fumon'in was Oguri Tadamasa's "official" grave.[127] Abe also
appropriated the authority of Ninagawa, who claimed that although there
were many Oguri graves, the one at Fumon'in was the most legitimate.

Arguments over the final resting place of Oguri's head exacerbated
the rivalry.[128] The search for it began in earnest when Nakajima Sanzaemon, a Gonda village official, returned to Gonda in 1869. He had escorted
Oguri's wife, mother, and daughter safely from Gunma to Aizu to Tokyo,
and afterwards worked as a police officer in Shizuoka Prefecture. According to the most popular account, Nakajima remembered his master's wish
to have head and body joined in death. Whatever the reason, Nakajima
planned a late-night dig in the Hōrinji cemetery, where Oguri's head was
buried. The first attempt failed. Nakajima and his conspirators then sought
help from the Takahashi Village headman, whose uncle worked in the local
government with jurisdiction over the Hōrinji Temple. The uncle scouted
the temple under the pretense of wanting to erect a gravestone for Oguri.
After successfully locating the head, the men sneaked into the temple
grounds at night, dug it up, and, according to the most reliable evidence,
brought it back to Gonda, where it was reburied on Mount Kannon.

Prefecture documents show that Oguri's and Mataichi's heads were
removed from the temple, but their final resting place is part of conflicting oral history. Gonda Village descendants of those who helped steal
the heads claimed Oguri's was returned to Gonda while Mataichi's was
reburied in Shimosaida Village, with the rest of his remains. In the mid-
1910s, a local researcher in Gunma conducted interviews with villagers
alive at the time of Oguri's death. They claimed that they had kept the
story about Oguri's head secret, fearing repercussions from the local government. When Hayakawa conducted the first research about the stolen
heads and published his findings in *JOJ* in 1923, controversy erupted
between Gonda and Shimosaida Village as two possible locations for the
heads, but the Fumon'in was not mentioned.[129] The first time the Fumon'in
became embroiled in the issue was in the 1930s, after Abe arrived. A 1935

JOJ article by Toyokuni argued that even though the Fumon'in had recently
tried to claim Oguri's head, "Gonda has superior evidence [of possessing
Oguri's head,] but people there are completely silent. . . . [W]e have no
choice but to advocate the Gonda theory for them."[130] An alternate story
regarding the whereabouts of Oguri's head, one retold by Abe supporters, argues that an Oguri retainer named Mukasa Ginnosuke stole the
head and brought it to the Fumon'in. However, for this story to be accurate, Oguri's head had to have been displayed on a pole after his execution, not taken to Tatebayashi for inspection, as was verified by several
sources.[131]

Regardless of the head's location, Abe stood to gain more publicity for
his temple by engaging in this controversy. He interviewed Hara Yasutarō,
one of the men in charge of Oguri's execution, for his 1941 Oguri biography. Abe clearly intended to use Hara's testimony to support the
Fumon'in's claim to Oguri's head. Leading his witness, Abe asked, "Did
you know that Oguri's head had been displayed on a pole, was stolen,
and then brought to his family grave?" Hara responded, "It's a fact that it
was stolen. I heard some rumor that it was taken to somewhere, maybe
Saitama, but I had no use for his head, so I just cut it off and then went to
Echigo to fight some rebels."[132] However, in an interview by another author in the early 1920s, Hara stated, "I saw the moment Oguri's head was
cut off," which has been understood to mean that he himself did not do
the beheading.[133] Either way, many thought Abe's interview with Hara
was bold because it asked pressing questions about the imperial army's
responsibility for Oguri's death. Novelist Ibuse Masuji was so impressed
by Abe's courage during what must have been an awkward interview for
both Abe and Hara that he published a short story in 1949 about Abe's
interview, titled *The Priest of the Fumon'in* (*Fumon'in san*).

Abe's persistence through the 1930s and 1940s paid off. His biography
received good reviews. Not surprisingly, one critic connected Abe's work
to an earlier work on another tragic Tokugawa figure: "I should say it is
even better than Shimada Saburō's biography of Ii Naosuke."[134] Eventually, the Fumon'in eclipsed Tōzenji as the premier Oguri site of memory
among national celebrities. Oguri's grave put the Fumon'in on the
map figuratively and literally—the Ōmiya City tourism bureau included
the Fumon'in on its map with the caption "This is also the site of Oguri's
grave."[135] Abe even created postcards depicting Oguri's meeting with a
former abbot at the gate to the temple. He succeeded in building his temple's reputation and received visits from various dignitaries—political,
military, academic, and artistic—all due to the temple's connection to
Oguri Tadamasa.

The temple, however, also attracted would-be treasure seekers, including one who claimed to be Oguri's grandson through an illicit affair. The ambiguity surrounding the nature of Oguri's death, rumors about his putative wealth, and lack of information about his personal life haunted his legacy in ways that did not afflict the legacies of the Aizu samurai or Ii Naosuke. Lacunae in the historical record encouraged invention, and Kawahara used this to forge a connection to Oguri. Treasure hunters continued to dig on Mount Akagi in Gunma Prefecture, especially during the 1930s, when there was a gold rush of such activity. According to Yoshio Kodama's *Sugamo Diary*, when Tōjō Hideki, Kishi Nobusuke, and Kodama were discussing rumors about two hundred million dollars' worth of platinum that had been thrown in Tokyo Bay, Kishi related it to the Akagi treasure: "Mr. [Prime Minister] Konoe's secretary, Ryūnosuke Gotō, once said there was some gold buried on Mt. Agaki. He dug for it, but found that there was no truth to that story either. Mr. Ikezaki commented, 'That fellow Gotō looked like the sort of man who would dig for gold that others had buried. He was just the type.' Everybody burst into laughter."[136]

Gotō seemed to have abandoned his gold fever before losing too much money, but others did not fare so well. In 1933, Inomata Kōzō, who became a prominent socialist politician after the war, defended his former schoolteacher in a fraud cause. The defendant, Seki Giichirō, had started digging at several sites, including Mount Akagi; an unknown location in Yamanashi Prefecture; and, after teaming up with a clairvoyant from Kobe, Komochi-yama, a mountain located halfway between the other two sites. He founded a group to help raise capital for his project, promising a return 220 times the initial investment of 10 yen, and eventually collected 35,000 yen from people as far away as Manchuria and Karafuto.[137]

But Kawahara Hidemori was singled out by memory activists for appropriating Oguri's name in his treasure hunting by claiming to be Oguri's grandson.[138] A local Gunma historian recounted Kawahara's treasure-hunting tale and how Kawahara had changed the face of the legend: "Dilettantes have been chasing the gold since the Taishō period, but people began flocking to Kōshū City, Yamanashi Prefecture (Kawahara's hometown) from the beginning of the Shōwa Period (mid 1920s)."[139] Toyokuni warned his readers that Kawahara was a fake, "the Oguri family stays away from him, and others who get to know him even a little bit will do the same."[140]

Just as Abe became famous through Oguri, so too did Kawahara, who often appeared in newspaper accounts about the Tokugawa buried-treasure story. Kawahara told Abe that he possessed a treasure map unearthed after the Kantō earthquake, and that if Abe would only help him find the treasure, Abe could have a new gate for the Fumon'in, while the rest would be donated to the navy. So convincing was Kawahara that even the ultranationalist leader Tōyama Mitsuru supported his cause. Kawahara and Tōyama requested a visitation to Oguri's gravesite at the Fumon'in. Claiming that he had no reason to reject their request, Abe allowed them to visit. "On the appointed day," wrote Abe, "journalists from nearly twenty newspapers showed up. Kawahara arrived wearing formal kimono with the Oguri family crest; this really irked me."[141] Abe noted the journalists' disappointment, however, when Tōyama did not attend, sending his son instead. Kawahara showed off his financial backing, paying three hundred yen for the funeral service and treating other attendees to fifty yen's worth of food each.[142]

Despite their competition over representing Oguri's legacy, Abe and his rivals in Gunma shared an intense dislike for those who used Oguri's image for selfish material gain. They agreed that there were improper ways to appropriate Oguri, in particular speculation about buried treasure. Oguri could only be discussed within the context of patriotism, loyalty to the state and the people, and a paragon of *bushidō*. During Kawahara's visit, Abe disavowed any connection between him and the hidden treasure during his sermon: "Integrity was Oguri's greatest quality. He also worked to Westernize the country's military installations. He was the epitome of Japanese *bushidō*. There are many things that future generations should learn from him and my job is to teach people this. Oguri had no material desire nor should he have. As for buried money, this mountain temple has nothing to do with it. What is needed more than money is an understanding of the spirit of the people. If I can impart this spiritual teaching to people then my job is done. I let people visit Oguri's grave so that they can learn about his greatness and this is why people should come."[143] Abe's speech fit within the context of the 1930s. Oguri's martial spirit, an example of *bushidō*, reflected the state's ideal for soldiers and citizens. Likewise, fascism attacked materiality, represented here by the search for Tokugawa gold. Abe transforms himself into a spiritual guide, filling in for Oguri's absence.

The Fumon'in's popularity did not survive WWII. Sadao grew suspicious of Abe's intentions regarding Oguri's legacy, and their relationship soured.[144] Abe lived long after the war, dying in 1978 at age eighty-four, but he did not promote Oguri as passionately as he had before the war. When I visited the Fumon'in in 2005, the priest told me that the last Oguri Festival at the temple was held when he was a child, during the 1940s.

When I pressed him for more details, he paused, then said, "Well, it's complicated." He also informed me that people rarely came to inquire about Oguri, but the temple's connection to Oguri is still listed in its brochure and website.

Conclusion

Oguri's legacy boomed from the 1920s to the 1940s through the interaction of local and national collective-memory structures. Oguri's historical memory gained salience during this time due to the creation of physical markers that anchored his past to the present. The central government often determined which values bestowed sites of memory with meaning, for example, commemoration through modern statue building and the bestowal of posthumous court rank. However, as the case of Ii's statue illustrates, local interests could push the government to clarify the use of space when erecting these monuments. Only after memory activists tried to erect Ii's statue in a Tokyo park did the government create a law to dominate public space in "national" cities. I do not suggest that cities are neatly separated into local and national ones; as Yokohama and Yokosuka demonstrate, they functioned as places for both local and national celebrations, bringing renewed legacy to Ii and Oguri respectively.

The stated goals of certain commemorative activities did not need to be successful to have meaning. Gunma supporters failed to obtain posthumous rank for Oguri, calling on the Shidankai, pro-Tokugawa historical essayists, and politicians to no avail. However, the discourse created through this activity strengthened Oguri's salience in local memory, further marking him as a tragic figure in Meiji Restoration history. Even the would-be treasure hunter Kawahara, who failed to gain legitimate recognition from anyone with a stake in Oguri's memory, still convinced people to lend him money for his ne'er-do-well adventures.

The Oguri boom experienced during the interwar period also illustrates the complex interaction between the local and the national. It demonstrates how local people internalized, momentarily at least, the nation. Gunma memory activists appropriated Oguri's valuation as a naval hero at a time when the navy helped define the military prowess inherent in the prewar and wartime concept of imperial Japan. Some historians relegate local interests to a passive role in the creation of the modern nation-state, national identity, and national history production, even though many scholars today acknowledge local ability to negotiate and participate in the variety of initiatives begun at the national level.[145] As far back as Yanagita Kunio, scholars have seen the countryside as a repository from which nationally recognized scholars pluck folktales, local heroes, and history to define Japanese identity. But as the cases of Ii and Oguri demonstrate, locally created discourse was present even in celebrations of the nation, as happened at the Yokohama and Yokosuka fiftieth anniversaries. Ninagawa's and Abe's works caught the attention of national celebrities, but even they borrowed heavily from local historians, often copying their narratives verbatim. Local people also appropriated master narratives of historical memory to construct their own regional identity that was unique to, but not in conflict with, a national whole. Thus, instead of seeing Oguri's legacy as the putative center of a local-national binary, it should be understood as the very network of local and national interaction.

CHAPTER 4

Re-creating Restoration Losers
in Postwar Japan

The war years dramatically shifted the focus of historical memory and commemoration in two ways. First, Japan's loss represented a defeat of the Meiji imperial dream, or at least the dream as it existed in the minds of nationalist and pro-emperor supporters. A new retelling of the Meiji Restoration and its primary actors occupied many postwar-period historical debates and made its way into popular culture that portrayed the Restoration. As academics, local historians, and writers moved away from the "great men" approach to history, their focus shifted from the heroes of the Restoration to those of humbler origins and the unnamed masses. Not all of the prewar heroes were forgotten; Saigō Takamori, for example, remained a staple in popular culture, but the values that defined heroism had changed. Support for the emperor was no longer the dominant narrative in the heroic life. Instead, postwar Restoration heroes were depicted as concerned about their own well-being first, while working for their community, peace, and occasionally the prosperity of Japan along the way.

Historical writing and popular culture have shared social and political contexts throughout the postwar era. Growing interest among historians in the commoner masses and low-ranking samurai, for example, paralleled similar developments in popular culture during the time of the student protests in the 1960s, with the resulting promotion of young (and often unmarried) men like Sakamoto Ryōma as the new heroes. A similar synergy occurred on a more modest scale for the Tokugawa "losers," as Japan itself had to deal with its own status as a loser. As this chapter will show, writers tried to redefine them within postwar historiography and popular culture with mixed success. During the 1950s, Ii Naosuke's popularity experienced a boom in the countryside, and Shiba Ryōtarō rewrote the Shinsengumi's story for young audiences in the 1960s. Oguri also appeared in postwar literature, through the works of Ibuse Masuji and Shiba Ryōtarō, and became the subject of a 1950s film.

Second, the government's efforts to revitalize the Japanese countryside gave memory activists new opportunities to project their own version of Restoration events onto the national scene. This process began as the structure of the countryside changed through the conglomeration of villages, towns, and cities during the 1950s, necessitating new collective memories to bolster new local identities. Memory activists benefited from the money that was being funneled from the central government to the countryside to promote the "hometown" (*furusato*) boom that started during the 1970s, and continued under campaign names such as "discover Japan," "exotic Japan," and the more recent "history road" (*rekishi kaidō*) of the 1990s. In the twenty-first century, memory projects have been bolstered by the numerous television programs that highlight obscure local history, products, and culture. This growth in domestic tourism was a journey backwards in time, usually into an Edo-period past, as much as it was a move away from an imagined urban center.

Although none of the modest postwar attention made Oguri a national hero, local memory activists did succeed in solidifying Oguri's story as a Gunma story, building on the foundations of its prewar predecessors and benefiting from the renewed attention to the countryside. Contests over the appropriation of Oguri continued as well when Kurata Village transformed into Kurabuchi Village during the 1950s. Postwar regionalism enhanced Oguri's image as a Gunma hero, which helped Kurabuchi activists eclipse their Saitama competitor, the Fumon'in, for recognition as the source of Oguri knowledge and the right to claim him as a local hero. But Kurabuchi still faced challenges from competing appropriations of Oguri, in particular those who focused on his connection to buried-treasure legends.

Locating Oguri in Postwar Historical Writing

In most postwar historical circles, the Meiji Restoration became a foundational issue toward explaining Japan's devolution into fascism, violence, and warfare. Under the influence of prominent leftist historians freed from the ideological constraints of wartime Japan, interpretations of the Restoration years coalesced into several schools of thought: a diverse group of Marxist historians, some with nationalist overtones in their work; modernists who tried to fit Japanese history within the framework of Western historical theory, and who looked for instances of individualism and democracy in early modern Japanese history; the optimistic modernization school that saw the course of industrialization, capitalism, and democracy momentarily interrupted by an aberrant militarist, perhaps fascist, movement; and folk historians (*minshūshi*) who found spontaneous intellectual, social, and cultural developments among "the people."

Narita Ryūichi argues that Tōyama Shigeki's 1951 *Meiji Restoration* (*Meiji ishin*) represented the dominant view of the Restoration in the early postwar period.[1] Tōyama, a Marxist historian, created what became one of the dominant narratives in Restoration historiography. He set the start of the Restoration period during the 1830s, when cracks within the feudal system began to appear—cracks that were exacerbated by Commodore Perry's arrival in Japanese waters. He focused on the convergence of commoner unrest with tensions between the shogunate and reform-minded *sonnō jōi* forces led by the Satchō domains. Tōyama contended, however, that even after the fall of the shogunate, feudal elements remained—Japan had not undergone a Marxist revolution.[2] The unique Japanese characteristics of modern Japan's semi-feudalism were defined, in part, by imperial absolutism. Other Marxist historians qualified Tōyama's basic description of the Restoration; Inoue Kiyoshi, for example, differed from Tōyama in arguing that the West threatened to colonize Japan. He and Ishii Takashi, in their respective works, blamed the shogunate, and Oguri along with it, for the resulting "semi-colonization" (*han shokuminchika*).[3]

Despite differing in intellectual agendas, many historians shared with Inoue and Ishii a tendency to portray both the old Satchō figures and the Tokugawa loyalists negatively. As Gluck noted, it was men like Sagara Sōzō, who had championed the commoners and promised that the new government would halve peasant tax burdens, who became the new heroes.[4] The shogunate's men had been part of the problem, often described as feudal and inept, unable to hold back the West, and the root of class conflict—they were never counted among "the people" lauded by folk historians. With Oguri's military and economic reforms now cast in a negative light, he was seen as pushing Tokugawa absolutism, for which, one author noted, "it is no surprise that Oguri was considered a criminal."[5]

In the postwar era, those who supported *bakumatsu* Tokugawa figures had to find ways to rehabilitate their heroes to make them palatable in the new historiography. Some writers did this by positioning their heroes against a shared, vilified institution: the emperor system. For example, Oka Shigeki's 1948 biography of Ii expressed anger toward the ultranationalism represented by the Mito rōnin who killed Ii and were subsequently valorized by receiving posthumous court rank.[6] He celebrated, through Ii, liberation from the emperor system experienced after the war, calling Japan's loss a "second opening" and a chance to reexamine Ii's value in American-Japanese relations.[7] Katsu Kaishū also was a convenient shogunate retainer for rehabilitation, because he could not be placed in any neat but problematic category such as Tokugawa versus Meiji loyalist. The Marxist historian Tamura Eitarō depicted Katsu as someone who was against the feudal system and desired an American-style government, but did not advocate ousting Westerners and was not an emperor zealot—criticisms leveled against Saigō Takamori. Tamura highlighted Katsu's commoner origins and the poor economic conditions facing low-class samurai like Katsu. Writing during the high economic growth of the 1960s, Tamura asserted, "People today cannot understand the difficult lives of these low ranking bushi."[8] Even better, according to Tamura, Katsu tried his best to avert warfare, believing that Japan could only unite if Japanese refrained from fighting one another.

This move to reverse the dynamic of prewar historical accounts also offered a forum for using Oguri to denounce the dominant narrative of the Meiji Restoration and the imperial institution, which put him, perhaps for the first time, in the historiographical mainstream. During the early 1950s, when he was well into his eighties, Ninagawa Arata published three historical works that continued his prewar condemnation of the pro-Meiji historical narrative. In *A Correct View of the Restoration* (*Ishin seikan*), Ninagawa attacks the Meiji legacy within the context of a democratic society under a new constitution. He argues that the Japanese should be awakened from their confused belief that the Meiji Restoration was a positive event and realize that it was, instead, "the action of conspirators, an action of great destruction."[9] According to Ninagawa's understanding, the fall of the shogunate had nothing to do with self-proclaimed loyalists. The last shogun ended the shogunate by voluntarily surrendering his title; the Restoration did not reflect the will of the emperor, nor did "the people" oppose the shogunate.[10]

Ninagawa promoted the Tokugawa legacy over that of the Meiji loyalists as the source of Japan's democracy. His emphasis on Yoshinobu's agency in Restoration history allowed him to de-center the role of the pre-WWII oligarchy. Imperial Japan had interrupted the transition to democracy initiated during the Tokugawa period. Similar to most prewar critics of the oligarchy, Ninagawa argued that the Meiji government used the Charter Oath to its own advantage and not to uplift the people. He conceded the democratic nature of the Charter Oath's first article: that "all matters should be decided by public discussion."[11] However, he contended that much of the Charter Oath, such as "the realm is not the realm of one person" and "the realm should be returned to the people," was based on earlier Tokugawa notions of democracy, what he called the "Tokugawa constitution."[12] Ninagawa suggested that Yoshinobu's resignation memorial informed the Charter Oath's concept of government for the people. The Charter Oath's article, "deliberate assemblies shall be widely established," copied Yoshinobu's "expand public discussion widely throughout the realm."[13] In other words, democracy was not a concept borrowed from abroad but had been a part of Japanese history.

In his second book, *Emperor: Who Is the Master of the Japanese People* (*Tennō: Dare ga Nihon minzoku no shujin de aru ka*), a best seller, Ninagawa historicizes the imperial institution and describes the emperor as a marginal figure in Japanese history.[14] For Ninagawa, the Meiji emperor system was the central anomaly that led to war. In the book, he cites early modern writings by Hayashi Razan, Asaka Gonzai, and Arai Hakuseki to support his view: "Feudal scholars freely critiqued the emperor," he argues, then quotes Hayashi Razan: "The emperor Kaika took his own mother-in-law as his wife, was that not unrighteous? A lord should behave within the bounds of humanity. How can a person of high status behave like an animal and still govern the country?"[15]

Ninagawa believed the relationship between the emperor and the people of Japan to be a tenuous one. The only emperors known to most Japanese were Kanmu, Tenji, Godaigo, and Meiji, and none, he said, could be called great men.[16] He posited the emperor position as a foreign concept, evidenced by the word for emperor—*tennō*—being of Chinese origin, yet conveniently ignored the foreign origins of many warrior-related terms.[17] The emperor was never considered the father of the Japanese people in Japan's two-thousand-year history, he argued, nor should a ruler-subject relationship exist in a democratic Japan.[18] Ninagawa even refused to accommodate the view that the emperor was merely a symbol of Japan, arguing that the emperor could not be considered royalty and was thus incomparable to European monarchs. In *A Correct View of the Restoration*, he wrote, "A country is a group of people united by their own sovereignty. The emperor is not a sovereign; [therefore] it is impossible for the emperor to unite the people."[19] It was dangerous, he argued, to call the emperor "a symbol of the people," as stated in the postwar constitution, because it reflected the continued presence of the Meiji imperial ideology in democratic Japan. Regarding the prominence of the emperor in the constitution, "we can say it's a plan to overrun democratic Japan by bringing evil views of the past into the present day."[20]

Ninagawa recast Oguri in light of these critiques while also attacking familiar heroes. Oguri appeared in both of the above works and was the focal point of a third biography, *A Pioneer of Opening Japan: Oguri Kōzukenosuke (Kaikoku no senkakusha: Oguri Kōzuke no Suke)*.[21] As in his previous writings, Ninagawa defined the Meiji period and its legacy as the bad "other" that had interrupted Japan's progress. He continued to target Katsu Kaishū and Saigō. Ninagawa was not alone in attacking Saigō; the postwar Marxist historian Tōyama Shigeki portrayed him as a feudal reactionary, as did Tamamuro Taijō, who described him as anti-modern and poorly educated, with "political self-righteousness driven into his bones."[22] In the context of a democratic postwar Japan, Ninagawa elevated Oguri's role in history to that of forerunner of democracy. He argued, "If Oguri's idea to create a centralized state had been used before the Meiji period then Japan would not have been forced to wait until now for democracy. Katsu did not have the know-how [to do this,] and Saigō was not even thinking about these things."[23]

At the time, Ninagawa's criticism of the emperor system was described as evidence that he underwent a "conversion" after the war.[24] At first glance, this observation seems logical; after all, he wholeheartedly supported Japan's encroachment into other countries. This reading of Ninagawa's work, however, conflates wartime patriotism with support of the emperor system and the Meiji legacy. Before the war, Ninagawa justified Japan's expansionist policies, but he never praised the Meiji emperor or the emperor system. Much of his postwar work builds on his earlier criticisms of not only the Meiji oligarchy, which was not uncommon at the time, but of the heroes and motivations of the Restoration itself. His postwar attack on emperor-centered ideology continued in this vein. Like many other historians, Ninagawa transformed his subject using the appropriate context of the era: speaking of a new, democratic Japan; configuring his hero to fit the discourse of Japanese nation and folk (*minzoku*); and avoiding state-oriented terminology, such as "state citizens" (*kokumin*), which evoked wartime ideology. His work accorded with postwar progressive historians, many of them Marxist, who saw the Restoration and post-1868 Japan as the root of Japan's wartime history. Although Ninagawa's critiques occasionally mirrored those of prominent leftist historians, many of whom, like Inoue Kiyoshi, also pointed out the imperial institution's recent creation, Ninagawa himself was not a Marxist, nor did he agree with their assumptions about the Restoration.[25]

Ultimately, and despite Ninagawa's many efforts, Oguri could not escape the generally negative portrayal of high-ranking shogunate retainers common in the evolution of postwar historiography. Protests against the Vietnam War and the Japan–U.S. security treaties intensified the trend among leftist historians toward an ethnic narrative driven by conflict with the West.[26] Ishii Takashi had augured such a trend when he claimed that the shogunate, under Oguri's direction, had facilitated Japan's "semi-colonization" through a putative deal with the French to exchange Hokkaido for assistance in fighting the shogunate's enemies.

Moreover, as Gluck has pointed out, a counter-narrative developed during the 1950s that extolled the positive values of the Restoration, depicting it as a peaceful transition toward modernity—a type of modernization theory that had existed before American historians of Japan exported it to Japan in the 1960s.[27] Ninagawa's efforts to use "feudalism" and its representative individuals, such as Oguri and Ii, as a positive counter to the Meiji oligarchy became anachronistic as the Restoration came to denote progress, and feudalism was blamed for the rise of prewar oligarchy, fascism, and militarism.[28] Postwar Japanese folk historians, who advanced a narrative that focused on the role of non-elite, Japanese folk, were hardly likely to celebrate a prominent official like Oguri, the very sort of archetype that folk historians hoped to free the Japanese people from during the postwar period.[29]

Popular Culture and Restoration Losers: *The Priest of Fumon'in Temple*

Literature and movies tried to address the civilian experience and expose the evils of war, with a large amount of the blame directed toward the Japanese state. This shifted during the late 1950s, when economic recovery changed the artistic climate, and Japanese themselves were gradually portrayed as victims.[30] Historical fiction and period films about the Meiji Restoration commented on war, loss, and hardship by drawing on the lives of Restoration losers. As "losers," their stories embodied loss in, and of, Japan: the empire, its status in the world, and the Japanese who lost loved ones and any sense of well-being.

Ii Naosuke was the first Restoration loser in the postwar period to undergo a reassessment in popular literature, kabuki, film, and eventually television. Like Oguri's rehabilitation, Ii's bridged the gap between national popular culture and regionalism. Death and war appear as twin themes in postwar fiction about Ii. Tanaka Hidemitsu's short historical works "Outside the Sakurada Gate" ("Sakurada mongai") and "Sakurada in Snow" ("Sakurada no yuki") focus not on Ii's assassination per se but on the lives of those who assassinated him. Tanaka, like Oka Shigeki in his biography of Ii, wrote from the perspective of his own wartime observations of death. Tanaka's treatment of the Mito rōnin reflected his shock over the suicide, in front of the emperor's palace, of fourteen men who belonged to the right-wing Daitōjuku. For Tanaka, their deaths, an extreme expression of remorse over Japan's loss, paralleled the Mito rōnin's murder of Ii; both were meaningless and born out of nihilism.[31]

Funabashi Sei'ichi's *The Life of a Flower* (*Hana no shōgai*) also approaches Ii's story through humble characters. Funabashi himself stated his desire to eschew the dark literature proliferating at the time.[32] In the book, Ii's spotlight is shared by his friend and vassal Nagano Shuzen, who falls in love with one of Ii's concubines. The story line shows a more personal, human side to Ii while slightly demonizing the overly violent pro-emperor faction that opposes him. As does the book, the subsequent kabuki version of *The Life of a Flower* minimizes the emotional distance between Ii and the audience; as one critic noted, "Unlike previous renditions of Ii, Sarunosuke's Ii is not overly arrogant."[33] In fact, another kabuki play about Ii, *Ii Tairō*, failed to become popular because it did not close the gap between Ii and the audience. The first few acts used Ii's life story as a lens to comment on society, but only late in the play does it begin to show his personal side—a narrative unevenness noted by at least one critic at the time.[34] Although a movie version of the book was also released in 1953, it was not until *The Life of a Flower* appeared in 1963 as NHK's first-ever Taiga drama that Ii's newly humanized image and popularity soared in the countryside. Only the first episode of the public television series still exists, but its impact on the national audience at the time was immense. In the year following its broadcast, nearly a million visitors descended on Hikone City, ground zero for what one reporter dubbed the "Life of a Flower boom."[35]

Although Ii flourished briefly in popular culture, Oguri's presence was less salient. But a few prestigious treatments in literature and film kept him from slipping into total obscurity. Ibuse Masuji wrote the first postwar fictionalized account about Oguri. *The Priest of Fumon'in* (*Fumon'in san*) retells Abe Dōzan's interview with Oguri's executioner, Hara Yasutarō. According to Ibuse, he first encountered Abe's story while working as a propagandist in the Japanese military in Singapore. His friend at the time, Abe's nephew, often talked about his uncle's exploits.

After the war, the nephew sent Ibuse a copy of Abe's biography of Oguri and asked if he would write something about his uncle.[36] Ibuse eventually visited Abe at the Fumon'in, a visit that is still a source of pride for the temple, as is evident on its website.[37] The story first appeared in 1949 in *Kaizō Bungei* and was reworked at least three more times, the last in 1988, five years before Ibuse died. Each successive version became shorter but included more about the connection between Oguri and Kurabuchi Village, reflecting the growing influence of memory activism in Gunma.

As several critics have suggested, *The Priest of Fumon'in* illustrates Ibuse's deep respect for Oguri's personality, and frustration with the attitude of "might makes right" (*kateba kangun*).[38] Oguri fit within one of Ibuse's favorite protagonist types—the marginal historical figure and the failures—as can be seen in his long-running serialized novel from the 1930s: *Waves: A War Diary* (*Sazanami gunki*).[39] The story follows an obscure but historical young warrior on the losing side of the twelfth-century Genpei War as he tries to survive against the victors. Marginal "losers" are also a theme in his castaway novel *John Manjirō*, which, like *The Priest of Fumon'in*, follows historical documents closely and resists simple fictionalization.[40] He believed, as did many Oguri memory activists, that an unspoken taboo had existed against writing about Oguri. Toward the end of his life, Ibuse commented, "Before the war, nobody wrote about Oguri, not even Kume Kunitake; the Ministry of Education put pressure on people [not to]."[41] *The Priest of Fumon'in* centers on the conversation between two old men whose conversation is weighed down by the chaotic events of Restoration history. Hara and Abe both fit the literary figure typically found throughout Ibuse's work: the eccentric old man, a figure Ibuse developed over his career.[42] Ibuse's verbatim retelling of Abe's meeting with Hara connects the twin themes of loss and age; as Katsumata Hiroshi points out, "Here you've got this green, clueless twenty year old working for the emperor's army, who cuts down a high-ranking bureaucrat, and this angry priest. I think Ibuse empathized with the priest."[43]

Ibuse starts by drawing readers into a dialogue between past and present. In the opening lines of the original story, the narrator states, "I saw history with these eyes. With these ears I heard the voices of people from history. History and the present mesh together like two teeth of a cogwheel." The story begins with the narrator telling a story about how he went looking for a suspicious monk who came begging at his home in Kamakura. At a nearby train station he finds a monk about the same age as the beggar and, believing he's involved in some kind of scam, follows him to the nearby residence of a member of the House of Peers. Having built up the suspense, the narrator listens in to the conversation by hiding outside of Hara's house, allowing Ibuse to let the interview unfold as it happens, capturing the awkward tension between the two men.

Like Ibuse's other postwar writings, *The Priest of Fumon'in* is a commentary on the war experience. Liman explains this interaction of past and present: "Like other writers of historical fiction before him, Ibuse turned to the historical theme to cast a most urgent, most personal contemporary theme into a classical context where it would acquire a depth of resonance, an epic breadth and a sheen of elegance that could hardly be matched in a contemporary work."[44] Echoing the Japanese populace's postwar anger at its government for causing the war, Abe's anger spills forth as he pelts Hara with questions about why he killed Oguri. As if interrogating a war criminal, Abe asks Hara if he regretted killing Oguri, why he did not try to consider the evidence against Oguri before executing him, and if he felt remorse for his actions. Hara, in a confessional tone, dutifully answers each of Abe's questions. Recalling the moments after Oguri's execution, Hara admits, "It was a momentary dream; there was neither love nor hate."[45]

The story does not judge Oguri's putative executioner. Abe's line of questioning is broken by a silent moment as the two share tea, after which he apologizes to Hara for attacking him. Ibuse, as in his other stories—particularly those about his wartime experiences—avoids forcing readers to make a Manichaean judgment about the character's behavior and instead pushes them to consider how it was that an ordinary person could do otherwise horrific things.[46] The story ends with Abe chanting on behalf of Hara's past sins, which Hara accepts.

Tanizaki Junichirō, who once visited Hara's residence as a schoolboy, expressed succinctly the ambivalence a reader might feel about Hara's actions during the Restoration. It is unclear how he learned about Hara's involvement in Oguri's death, whether through Abe's 1942 book or Ibuse's short story. He only stated his surprise, after reading "a story about the Restoration," that Hara had led the group that arrested and killed Oguri, adding, "I'm an Edo native, but have no particular love or hate for the Tokugawa shogunate. And even now, having learned about Hara's past, I don't have much to say about it. But if I knew about Hara back then, well, I wouldn't have felt good about it. And for an Edo native to bow down and benefit from a Chōshū man who killed a Tokugawa retainer is pretty gutless—I feel such shame."[47]

Later versions of the story demonstrate how the center of memory activism shifted away from the Fumon'in to the Tōzenji in Gunma. Ibuse visited Gonda in 1951 to interview Ikeda Kayo, a woman who spoke to Oguri when she was seventeen. Ibuse recalled in 1985, "Years ago I went to visit the priest himself [Abe]. . . . I also went to Oguri Kōzukenosuke's grave deep in the woods from the Karasu River to hear the opposite side of the story. After gathering some documents I returned home and completely rewrote the story."[48] In the second version, Ibuse eliminated the leisurely paced introduction of the first. There is no narrator letting the story unfold, only a brief introduction about Oguri and the Fumon'in, followed by the conversation between Abe and Hara. But the story ends with a new section explaining the Gonda villagers' roles in helping the Oguri women escape to Aizu, an episode completely absent from the original story. Ibuse's last rewrite, published in 1988, continued to shift the attention from a focus on Abe Dōzan's story to Kurabuchi and the Tōzenji.[49] Ibuse only met Abe Dōzan once, between the second and third versions of *The Priest of Fumon'in*, and only a year before Abe's death.[50] The meeting did little to impress upon Ibuse any sense of a deep connection between Oguri and the Fumon'in itself; however, the gradual inclusion of the Gonda/Kurabuchi side coincides with greater memory activism in Gunma and Ibuse's unwitting participation in it. The change in Ibuse's short story over time from the immediate postwar period to the Heisei era (1989–present) demonstrates how local memory activism dominated the national presentation of Oguri—in this case, through short-story narrative.

The Birth of Tokyo: Restoration Losers in Postwar Period Films

As in literature, postwar historical period films (*jidaigeki*) retold the Meiji Restoration with one eye kept on WWII. The postwar-occupation-government censored movies forbidding most themes found in period films: feudal loyalty, honorable suicide, brutal violence, anti-foreign nationalism, and militarism. Although this did not completely eliminate the period-film genre from production, it severely curtailed it, precluding classic tales that celebrated almost all of the above prohibitions, including the story of the forty-seven rōnin.[51] Filmmakers avoided censorship through self-editing by, for example, having a protagonist or villain captured when a climactic fight scene would have been expected. Foregrounding pro-occupation values also worked as a strategy: the 1946 film *Kunisada Chūji* depicted Kunisada, a local gangster in *bakumatsu* Gunma, as a pseudo-union leader.[52] On the other hand, even seemingly harmless stories could encounter resistance from occupation censors, such as a movie about the relationship between Townsend Harris and Ii Naosuke. The government claimed that the production company needed to clear the project with the Harris family, and censors did not like the anti-Western portrayal of Okichi, the woman sent to "serve" Harris.[53]

It was not until after the occupation interlude that period films could return to serving, much like historical fiction, as a space for producers and writers to comment on contemporary issues through the lens of historical events. This had been true even before WWII, when the genre developed during Japan's most politically and economically troubled decade—the 1920s through the late 1930s.[54] While some producers of the era depicted scenes of mass peasant rebellion or warfare as reflections of contemporary invasions or labor unrest, others tried to foster an image of Japanese oneness. To this end, Restoration losers frequently appeared in prewar period films to demonstrate that even historical foes could cooperate. Such was the case in the 1938 *Numazu Military Academy* (*Numazu heigakko*), in which a former Chōshū samurai acts as a bridge between the oligarchy and a military academy dominated by ex-retainers. According to the film's narrative, both groups worked, despite their differences, to improve Japan against aggressive Western powers, namely America and Great Britain.[55]

Losers featured heavily in postwar period films, as they had in historical fiction, as stand-ins for Japan's loss and suffering during and after WWII. Ōsone Tatsuyasu directed other "loser" films, such as the movie version of Funabashi's *Life of a Flower* (*Hana no shōgai*) and *Rats of the Town* (*Edo no konezumi tachi*), about poor samurai living in Edo who are forced to turn to thievery and gambling after the Satchō armies enter the city, a commentary about the poverty and suffering experienced during the immediate postwar occupation period.[56] Films by other directors concerned with Restoration losers include *Edo Sunset* (*Edo no yūbae*), about survivors of the Boshin War in Hakodate who return to Edo, and *Samurai of the Great Earth* (*Daichi no samurai*), which also traces the hardships of Restoration losers—in this case, samurai from the northeast domains who move to Hokkaido and become farmers.[57]

As in postwar literature, few viewers would have been familiar with Oguri, and as such, he appeared only briefly in films throughout the 1950s. He is almost always portrayed as a villain or fall guy. In *Tsubanari Ronin*, Oguri is an elder councilor (*rōjū*)—a position he could never have achieved as a non-daimyo—in charge of retrieving a stolen document revealing the shogunate's plan to buy new weapons from a Western country. The deal is seen as unfavorable to Japan, and Oguri hires a rōnin to keep the document from falling into the hands of Satsuma men, a task that the rōnin ultimately fails to accomplish.[58] In the same year, Oguri was depicted in virtually the same role in the fourth installment of the Black Hood series (*Gozonji kaiketsu kurozukin: shinshutsu kibotsu*). In it, Oguri and a few other shogunate officials are watching the demonstration of a flame-throwing tank, reminiscent of the fire images of the bomb, given to the shogunate shortly after it has signed a commercial treaty with Prussia. The protagonist, a black-hooded rōnin who rights wrongs with two pistols at his side, shows up on a white horse and destroys the weapon, much to Oguri's dismay. The Prussians do not have another tank for the shogunate, but Gelda, the half-Japanese, half-Prussian, daughter of its inventor, agrees to give the rights to the tank design to whoever can find the blueprints. The blueprints are on a vase that is making its way to a Confucian scholar who befriended Gelda's father, and the Black Hood is out to protect him. Oguri hires a band of rōnin to track down the vase and kill the hero. In the climactic fight scene, Gelda saves the hero by jumping in front of gunshots. As she lies dying in his arms, she expresses her hope that Japan will be a peaceful country, where people will no longer kill one another. The Black Hood responds: "To create such a Japan, we will fight and die. We will build a peaceful Japan to give meaning to the many who have died."

Both films are largely forgotten among the many postwar period movies, but their very marginality suggests two ways in which WWII themes, and Oguri memory, had become commonplace within the genre. First, unlike other period films with darker plot lines that connect, quite overtly, Restoration losers' experience to that of WWII, these two films focus on the action without developing a complex plot. The Black Hood series is especially campy and was probably intended for a young male audience. Each film opens similarly: the shogunate signs commercial treaties with Western countries that also include military cooperation and the purchase of "advanced weapons," no doubt a commentary on the U.S.-Japan security treaty formed at the end of the occupation. Each opening scene also assumes that the agreement, signed by an undemocratic regime, is disadvantageous to the people of Japan, and that only someone from the margins of society can save the country. In both films, Oguri, although only a minor character in the opening act, is the villain who sets the two groups of protagonists against each other. In the first film, Oguri is depicted as a senior councilor. Why did the filmmakers not choose a more famous historical senior councilor, such as Andō Nobumasa, Ogasawara Nagamichi, or Mizuno Tadakiyo—men who served the shogunate during the 1860s, when each film takes place? Perhaps Oguri's villainous reputation was more familiar to postwar audiences. In any event, those who created the first movie to feature Oguri as a main character consciously chose an interpretation of his history that contradicted what they believed to be mainstream criticisms of him.

In 1958, Shōchiku studios released *The Birth of Tokyo, the Bell of Edo* (also called *The Birth of Tokyo*), directed by Ōsone Tatsuo, to celebrate Shōchiku's thirty-fifth anniversary of producing period films.[59] Today, *The Birth of Tokyo* is virtually unknown, receiving little attention in Japanese surveys and no coverage in English-language surveys. It has been mistakenly classified as a *chambara* film, a subgenre that refers to movies featuring swashbuckling violence, of which there is little in *The Birth of Tokyo*.[60] Nonetheless, it is significant because Shōchiku chose to mark its anniversary with Oguri, the film's tragic hero, during a transitional historical moment when other directors were commenting about WWII and its aftermath within their own genres.

Ōsone wanted to a make a movie about the Meiji Restoration from the loser's point of view. "Five years ago, films were still popular that purported to tell true events of the Meiji Restoration. We wanted to focus on the life of Oguri Kōzukenosuke, a man who was killed by the victors of the Restoration who wrote the Restoration's official history."[61] Ōsone and the producer, Kishimoto Gin'ichi, first contacted Ninagawa, who had just published his most recent Oguri biography. "We contacted Ninagawa, but when we spoke with him, he said the Meiji emperor was a young fool, Saigō was politically power mad, and Katsu was a politician only good at getting ahead in the world. Oguri alone was looking out for the good of the nation."[62] Both men felt that Ninagawa's interpretation drifted too far from the mainstream understanding of the Restoration and looked elsewhere for information. "There was even a woman in Itabashi ward, Tokyo, who claimed to be Oguri's granddaughter."[63] In the end, Kishimoto and Ōsone decided to avoid relying on personal informants for Oguri's story, opting instead to read everything they could about him. By 1957, they had decided on the following story line.

The movie begins with the Tokugawa shogunate's defeat at Toba-Fushimi and climaxes with Oguri's removal from office. The plot centers on Oguri and his supporters, who try to convince the shogun to fight against the advancing Satchō forces, versus Katsu Kaishū, who leads the call to capitulate. Oguri and Katsu represent the two choices available to Tokugawa retainers and the film's audience, while the subordinate characters—Oguri's adopted son, Mataichi, and a woman named Oryū—represent the emotional consequences of those choices. Caught in between the political struggle, Mataichi takes Katsu's position because he believes that this will prevent civil war. This strains his relationship with his father, who disowns him, and causes problems between Mataichi and his colleagues, who are mostly Tokugawa diehards, including the father of his fiancée, Saiko. Opposite Mataichi's character is Oryū, a woman who works in a "boat lodge" (*funayado*), ferries Oguri through

Edo, and secretly falls in love with him. She vehemently dislikes Katsu and his trust in the Satchō samurai. A flashback scene later reveals that a Satchō soldier killed Oryū's parents and raped her, and thus she fears the violence that the Satchō forces will bring to the city if the Tokugawa shogunate collapses.

The movie climaxes during a meeting between the Tokugawa retainers and the shogun. The shogun announces he will not fight against Satchō, causing great distress within the prowar faction. As the shogun rises to leave the meeting, Oguri grabs the shogun's *hakama* in a last-ditch effort to convince him to fight. Oguri is fired, men start to cry, and several retainers rush out into the courtyard to commit *seppuku*, thus ending the first half of the movie.

In the second half, Oguri and his pregnant wife leave Edo for what is supposed to be a quiet life in Gonda Village, while the emperor's forces march north into Edo. The imperial forces enter Edo, causing havoc, and during one bout of drunken debauchery among the imperial soldiers, Oryū finds the man who raped her years earlier and kills him. Katsu intervenes on her behalf and convinces Saigō to spare her. He also attempts a plea for Oguri's life, but he is too late—Oguri and his retainers have already been arrested and executed in Gonda. His father's execution traumatizes Mataichi, who abandons his earlier pro-peace position and joins the Tokugawa holdouts in Hakodate. They are quickly beaten but are pardoned, and Mataichi rejoins Saiko at a battlefield hospital, where she serves as a nurse in the newly created Red Cross. The movie ends with two juxtaposed scenes. In one, Oryū plants a seedling in front of Oguri's grave in Gonda as the sun sets over Mount Haruna. This is a common ending in period films; a woman survives to pray for the tragic hero's soul.[64] In the other, Mataichi, Saiko, and Katsu are back in Edo watching fireworks. Katsu's final hesitant but optimistic remark echoes a sentiment found in much postwar artistic expression: "Many lives have been sacrificed in making this [the new Japan] happen." Out in the streets, people dance, and an old man, confused by the commotion, is told by a young woman that the emperor is in the city, which is now called Tokyo. Awed, the old man says, "It's like a dream."

The Birth of Tokyo features many of the classic elements found in the period-film genre. Like many postwar samurai films, the central problem in *The Birth of Tokyo* is the question of loyalty, especially, as David Desser puts it, "loyalty to a society which may be in the wrong."[65] According to the film's narrative, neither the feudal Tokugawa shogunate nor the vengeful Satchō-led armies hold the solution to society's problems. Loyalty itself is the problem, as depicted during the two juxtaposed opening scenes in the film. The first shows Yoshinobu surrounded by his top retainers on the deck of the *Kaiyō Maru* as it sails back to Edo after an embarrassing defeat at the battle of Toba-Fushimi. Yoshinobu's men implore him to continue fighting the arrogant Satchō armies that dare insult and threaten the shogunate's three-hundred-year rule. But he demurs: "If I have to choose between the Tokugawa and Kyoto, I must choose the imperial court. Nothing can be done about it; it is the way of our times. I will retreat quietly to Edo." The scene then breaks to Saigō Takamori and his supporters, who kneel in front of the Meiji emperor and make a similar plea to their master: "Yoshinobu is an enemy of the court; we must cut down such enemies." But the Meiji emperor rejects violence: "Saigō, do not forget that the purpose of this political conflict is to create a new Japan."

Read as a metaphor for postwar Japan, Edo-turned-Tokyo represents the new Japan, emerging from a nightmare brought home by a defeated army as it prepares for the expected horrors of an occupation army. In the film, as in real life, the feared violence never comes; the mood is cautious but positive. Still, the film's portrayal of an occupation army penetrating into conquered land highlights the sexual anxieties of male "losers," whether Tokugawa retainers or postwar masculine pride. An Edo official tells Katsu he will prepare the women of Yoshiwara to stave off possible violence, an all-too-real reflection of comfort stations set up by the postwar government for the same purpose. The film has a moment of catharsis when Oryū finds the man who raped her, stumbling drunk out of an inn during a night of drinking, and stabs him.

Viewed as a commentary on the war, the film suggests that blame cannot be affixed to the leadership but on the second-tier military officers who fought in the leader's name. This was the narrative being told about WWII: a powerless Hirohito could not stop the war machine controlled by those who fought in his name. Removing Hirohito from responsibility for the war saved the Japanese people; he became an almost Christlike figure, except that he is kept alive to wipe away their sins and resurrected as a symbol of the nation. In one intimate scene between Katsu and Oguri, as they ride a riverboat to their respective homes, Oguri asks Katsu to take care of his son, Mataichi. Oguri reveals that his true feelings are the same as Katsu's, but due to his position within the shogunate, he has no choice but to press for war. Oguri's actions, like those of many within the wartime government, are excused as "doing one's job." The Satchō army is the clearest enemy in the film. Saigō, in particular, is portrayed as a bloodthirsty character who wants to punish the shogun for being an enemy of the court. While Oguri claims that he fights to protect Edo citizens,

Saigō is depicted as fighting only out of emotion. Even the British diplomat Alcock refuses to help Saigō, believing the Satchō invasion of Edo to be an offense against "international law and morality."

Oguri has a vision for Japan's future but is bound by loyalty to a shogun who is unwilling to defend the shogunate. His struggle over duty versus personal feelings (*giri ninjō*) makes Oguri the ideal period-film protagonist, embodying the "nobility of failure"—one who chases after a failed cause, never gives up, and dies in the end.[66] The film does not primarily cast Oguri as a defender of the shogunate but as a protector of Edo residents. While Katsu tells fellow Tokugawa retainers that he is glad Yoshinobu lost during the battle of Toba-Fushimi as a first step to avoid war, Oguri pleads with the elder councilors to defend Edo and "create a way for citizens to live to their fullest." Throughout the film, Oguri repeats that he is fighting for the citizens, a subtle but significant difference from prewar depictions of him as someone fighting for the "countrymen" (*kokumin*), a recurring meme during wartime Japan. Although individuality trumped the group as the focus of postwar literature and film, there was still an emphasis on the individual's role in caring for the local community.[67]

The film can also be understood as a critique of Japan's treatment by the victors. As one commentator noted, Oguri's execution just for advocating war fits into the "might makes right (*kateba kangun*)" formula, in which the victors can ignore their own actions and punish others. In this way, the film acts as a general attack on the irrationality and violence of war.[68]

Despite its all-star cast and large production budget, *The Birth of Tokyo* failed to become popular. One critic noted a lack of depth, poor supporting character development, and the depiction of women who were too idealistic (*tadashi sugiru*). Narrative problems also hindered the movie. In one scene, for example, Oguri tells Edo officials that they should burn down the city if the Satchō troops don't enter peacefully. This counters the director's attempt to portray Oguri as Edo's protector. More to the point, one reviewer stated it simply made no sense.[69] Nor did the movie follow typical trends in most successful postwar period literature and film. First, there was too much distance between Oguri, a high-ranking retainer, and the audience; they could not relate to his high status. This is not to say, however, that all elite historical figures were unpopular in postwar popular culture. Yoshitsune, perhaps Japan's most famous "loser," continued to engage audiences because he tried to protect his kinship group, the Minamoto, by refusing to resist his older brother Yoritomo's efforts to eliminate him. *The Birth of Tokyo* tries to depict Oguri as a defender of local citizens, but that is overshadowed by his defense of the shogunate. Ultimately, the most popular postwar Tokugawa-period figures, historical and fictional, were low-ranking samurai or commoners who became strong, average people with whom the audience could identify.[70] Men such as Yoshida Shōin, Ōshio Heihachirō, Sakamoto Ryōma, and the Shinsengumi members were celebrated in historical works of the 1950s and 1960s as young men who died in worthy causes leading up to the Restoration. Sakamoto Ryōma and members of the Shinsengumi—in particular one of its leaders, Hijikata Toshizō—were resurrected in the works of Shiba Ryōtarō.

Shiba Ryōtarō: The Restoration Hero Maker

It is no exaggeration to say that starting in the 1960s, no novelist has had a greater impact on the public's imagination of the Meiji Restoration than Shiba Ryōtarō. His historical writings have had tremendous influence on popular views of Restoration historical figures, in particular Sakamoto Ryōma and members of the Shinsengumi. What makes Shiba's work interesting in the context of Oguri's commemoration is that his view changed over time, from a negative one (covered in this chapter) to a positive one, as the work of memory activists increased during the 1980s and onward (chapter 5)—a trend also reflected in Ibuse Masuji's work.

Narita Ryūichi calls Shiba the most popular writer of postwar Japan because of the range of his work, the variety of topics he addressed in journals of different genres, and his wide readership. For Narita, Shiba was a "writer of the people" because he dealt with the "nature of Japan" (*kuni no katachi*).[71] Shiba's attitude toward history was motivated in much the same way as that of "people's" historians such as Irokawa Daikichi, men born during the first decade of the Shōwa period who, according to Gluck, traced their interest in modern history to the same formative events, WWII being the most prominent. Too young to have participated in the fighting, these men felt betrayal rather than guilt.[72] And, like many of those men, Shiba died in the mid-1990s, during the anniversary of the end of WWII, when postwar optimism was giving way to the economic recession, natural and manmade disasters, and political scandals. Shiba's work, and death, Narita argues, was coterminous with that of postwar (*sengo*) Japan.

As Keene once remarked, many of Shiba's historical works on Japan cover chaotic times, either the period leading up to the founding of the shogunate or its final years.[73] Shiba's writing on the Restoration years typically features heroes from marginal backgrounds: low-ranking samurai like Katsu Kaishū, Sakamoto Ryōma, Hijikata Toshizō, and Kawai

Tsugunosuke. He even portrays the last shogun, Tokugawa Yoshinobu, as an outsider: born into a much-maligned cadet family, he did all that he could to avoid becoming shogun, and even when he accepted the position, he knew that it would end badly for him.[74] The marginality of these individuals also characterizes their attitude toward their times; they approach the chaos in a rational way, avoiding the fanatical and mystical, and sometimes avoiding war.

Shiba's historical novel about Sakamoto Ryōma, *There Goes Ryōma* (*Ryōma ga yuku*), serialized in *Sankei shinbun* from 1962 to 1966, reflects the political and economic context of the 1960s era of high economic growth. Shiba portrays Sakamoto as an economic thinker who understood the value of money and commerce. Money was depicted negatively in most historical novels prior to this time, usually associated with villains, but Sakamoto is told he should engage in commerce because it is the future of Japan.[75] Part of what attracted Shiba to historical figures of humble origin was their noncommittal political views. Shiba disassociates Sakamoto from the extremists in his native Tosa domain—the fanatical supporters of the emperor—a gesture representing Shiba's postwar feelings about the imperial institution in contemporary history.[76] Shiba's Sakamoto Ryōma is less concerned with emperor worship and support for Satchō than with his own attempt to help modernize Japan. *There Goes Ryōma* also contains many characters who died during the Restoration; Shiba's message to a 1960s audience, Narita argues, was that they should not forget those who died during the war.[77]

On the losing side of the Restoration, Shiba's portrayal of the Shinsengumi in *Burn, Sword* (*Moeyoken*) highlights his desire to distance himself from political events of the 1960s, such as the student protests against the Japan–U.S. Joint Security Treaty.[78] Shiba wants to explore how a man should live during chaotic times; as in his other works, humble men dominate.[79] The narrative begins with descriptions of peasant swordsmen in the Kantō region, the type filling the Shinsengumi membership, most importantly, Hijikata Toshizō and Kondō Isami. The protagonist, the Shinsengumi co-leader Hijikata Toshizō, is not interested in the broader political events of the country. Throughout the story Hijikata differentiates himself from co-leader Kondō, who, Hijikata often complains, is too caught up in the politics of the times.[80] Hijikata describes himself as a "brawler," interested only in fighting. Just as an artist's goal is to create art for the sake of art, Shiba states, so too does Hijikata fight simply because he is a fighter.[81]

Shiba and other writers agreed with the many historians who faulted Oguri for opening Japan up to semi-colonization. Shiba believed that the outer domains abandoned the shogunate because it depended on the French for financial and military support: the conventional first step European powers used to establish colonies. As the primary advocate of French support, Oguri was the guiltiest among shogunal retainers in that he endangered Japan's independence, an argument Shiba makes in *There Goes Ryōma*.[82] Kaionji Chōgorō, writing about Oguri in 1968 against the backdrop of the Vietnam War, was even more forceful on this point. He attributed all of Oguri's knowledge to Roches, and stated that if there were people like Oguri on the Satchō side who were dependent on Great Britain, as Oguri was on the French, then Japan would have been split into two like Korea and Vietnam.[83]

In *The Eleventh Shishi* (*Jūichibanme no shishi*), Shiba dismantles Oguri's reputation as a stoic bureaucrat. Like Shiba's other works during the 1960s, the story follows the life of a marginal figure, a fictional character named Tendō Shinsuke, a peasant swordsman from Chōshū who is ordered to kill Oguri. Tendō initiates his plot by sending his adopted sister, Osae, to Edo to ingratiate herself with Oguri. Osae is torn; she has been sent to spy on Oguri, but she is attracted to him. Shiba uses her internal dialogue to describe details about Oguri's background, including Kurimoto's story about Oguri intending Yokosuka to be a "house with a storehouse attached," which is depicted as a project created for the shogunate's honor rather than out of patriotism. Osae eventually succumbs to Oguri's seduction: "She told herself she was doing this for Chōshū ... [after all,] Oguri sold the country to save the shogunate ... for the court, for Chōshū, for everything, she was just trying to give herself a reason. ... [T]here is no real Osae, just a body with a name attached, this body with the name Osae will test Oguri."[84] They have sex, and when Tendō arrives in Edo to meet her and kill Oguri, she's unapologetic about losing her virginity to Oguri. Shiba expresses his admiration for Oguri but ultimately portrays Oguri as villainous, cold, and calculating. Osae offers her body for the greater cause, just as Oguri offers up the country to save the shogunate. But the reader is left wondering if the loss of innocence in either case is worth it.

The only Tokugawa retainers worthy of praise, in Shiba's view, were low-ranked Meiji-era heroes, such as Katsu Kaishū and Fukuzawa Yukichi. During a 1978 interview between Shiba and comparative-literature scholar Haga Tōru, the men discussed the 1860 embassy and its impact on the men who participated. Those who visited America, Haga noted, returned to Japan with the intention of changing the country. Shiba responded, "However, Oguri Kōzukenosuke was also a member of the embassy, and even though he saw many things, one needs a megaphone to spread those ideas. Katsu and Fukuzawa had that ability." Haga concurred: "Correct, men like Muragaki Awaji no kami wrote interesting diaries . . . but they could not act as loud speakers. . . . Katsu's presence on the embassy was crucial."[85]

Marius Jansen notes that before *There Goes Ryōma*, "outside of Tosa, where he was part of local lore, Sakamoto was relatively little spoken of; his image was that of a swashbuckler and a stormy petrel whose early death prevented him from exercising much influence on the modern state."[86] Shiba brought Sakamoto back from obscurity and shaped him for a postwar audience, as he did for others like Hijikata Toshizō and Kawai Tsugunosuke. When writing about Sakamoto, Shiba relied on prewar writings by local historians. There is no evidence that he did the same for his portrayal of Oguri, about whom he seemed largely disinterested in using as a main character. But when Oguri did appear in a more positive light in Shiba's writings, it was only after decades of intensive local memory activism in Gunma, which resurrected Oguri as an internationalist and modern economist.

Local Boosterism and the Commemoration of Oguri

While Oguri appeared in small but nontrivial ways in film and popular literature during the 1950s and 1960s, his reputation in the countryside was enjoying a resurgence thanks to government policies that attempted to revitalize the countryside. When Ubuchi and Kurata villages merged to form Kurabuchi Village in 1955, Oguri functioned as an important source for unifying village identity. Tōzenji, which once dominated all local commemoration of Oguri, and an Oguri memorial group handed over leadership of memory activism to the newly formed Kurabuchi Village administration. In return for formal support and greater access to resources, village officials received a compelling foundation for local village identity as well as the potential to increase tourism.[87] Since the 1960s, locals in Gunma have succeeded in bringing national attention to Oguri and to their own role in his legacy. The novelist Ikenami Shōtarō recalls being sent out to Kurabuchi in 1962 to do research about Oguri. While working for NHK, he received a request from the Maebashi City branch to write a radio drama for its Gunma audience. Ikenami was less than thrilled: "For no reason in particular I just did not like Oguri . . . but I thought that it wasn't right to hate someone without doing a bit of checking, and I wasn't completely uninterested in him, so I took the job."[88] When he arrived in Kurabuchi, the mayor and local historians had nothing but praise for Oguri, and Ikenami had expected as much:

"It's natural for local people to praise local heroes."[89] But Ikenami became a convert and included Oguri in his later book *The Warring States and Bakumatsu: Men of Troubled Times* (*Sengoku to bakumatsu: ransei no otokotachi*).

By the 1970s, the memory activists of Kurabuchi Village and Gunma Prefecture had established Oguri as a regional figure. Their efforts coincided with the rise in the national discourse of "native place" (*furusato*) as a reaction against the rapid industrialization, urbanization, and westernization of postwar Japan.[90] The making of native places (*furusato zukuri*) allowed local people to solve problems that arose from rural depopulation by playing on Japanese "nostalgia for nostalgia."[91] *Furusato* and tourism did not begin at the local level. Marilyn Ivy's study demonstrates how Japan National Railways' highly successful advertising campaigns of the 1970s and 1980s used these images, but they can also be found in popular culture and even in the academic discourse of folklore studies.[92] However, both local and national interests, such as the ruling Liberal Democratic Party, shared in the celebratory *furusato* themes: the beauty of nature, compassion, nostalgia, motherly love, local dialect, and local history.[93] Oguri supporters used these developments as a new context for articulating their appropriation of his legend.

Oguri memory activists had experienced a series of devastating losses during the war years. In 1937, much of Tōzenji burned down during a fire that spread throughout the village, destroying the few Oguri relics not already scattered by Meiji troops nearly fifty years earlier. The Oguri family itself was in danger of extinction. Sadao and Kuniko's first child died young, leaving only one child, Mataichi, to carry on the family. He tried to make a living in journalism, much like his father and uncle, Yano Fumio, but repeated failure forced him to live off of Sadao's inheritance. After a bout of depression, he committed suicide by poison in front of his grandfather's grave in Tokyo; he left no suicide note, but the place of death spoke to the pressure he felt trying to live up to the Oguri legacy.[94]

For the remaining Oguri family members, however, it was their connection to memory activists that helped them persevere through the difficult war years. When American planes firebombed Tokyo in 1945, Oguri Kikuko, along with her mother and two children, Yoko and Tadahito, took refuge in Gonda. From the end of May until February 27, 1946, the family lived in a makeshift shack in front of the Tōzenji that had been owned and refurbished by the local agricultural society. Kurata and Ubuchi village officials provided additional support, and the children attended a local school.[95] The Oguri family eventually returned to Tokyo;

Tadahito continued to participate in Oguri memory activism, attending events in Gunma, and contributing to locally produced journals.

The cultural rebuilding of Gunma identity immediately after the war also included Oguri but faced opposition from occupation authorities. In 1947, the newly established Gunma Cultural Association, under the direction of Urano Masahiko, created the first Japanese-style card game to promote local history and identity among children at a time when history education had been suspended by the occupation. Called Jōmō Karuta, the play follows the traditional card game One Hundred Poets, One Poem Each (Hyakunin Isshu), except that these cards feature Gunma prefectural sites, geography, products, events, and historical figures. Urano charged a committee of eighteen local historians and educators, among them the former *JOJ* editor Toyokuni Kakudō, with the task of gathering materials for the project. Committee members submitted possible ideas for cards, totaling forty-four, including ones featuring Oguri; Kunisada Chūji, a local Robin Hood–styled bandit hero; and Takayama Hikokurō, a pro-emperor samurai. Oguri's card read, "Oguri, deeply insightful, he was falsely accused." The card proposals were organized by category and voted on. Oguri, along with Gunma native and Christian intellectual Uchimura Kanzō, scored low on the popularity scale, but Urano was an Oguri fan and ensured that Oguri was included in the deck. The initial proposal did not pass the occupation censors, who rejected the following cards, stating, "Oguri was a samurai, Kunisada a yakuza, and Takayama a supporter of the emperor."[96] Years later, Urano told his daughter about how he tried to convince a young GI to allow Oguri in the deck, arguing, "It is true that Oguri was not a Gunma native, but he was a lord here, was caught up in the trouble here, and killed. This is where he's buried, and the natives here yearn for him. He brought back ideas from America, created Yokosuka, and had a patriotic heart. He was a model *bushi*. As an American youth, as someone from a country of independence, you should ask yourself what you would do in his place."[97]

Ultimately, fearing that the cards might not pass the censors, he withdrew the three men. Feeling that there nonetheless should be some way to acknowledge them, he wrote the card beginning with the syllable "ra" to take their place: "lightning and dry winds (rai *to karakkaze*) obligation and humanity (*giri ninjo*)." As a last point of resistance, instead of placing the cards in the deck in the typical "i, ro, ha" order, he placed "i" first, then "ra," which would normally appear in the middle of the deck, and highlighted both in red. These were followed by the cards for the three Gunma mountains "because these three men towered in their contributions to Gunma."[98]

Despite the barriers to Oguri commemoration at the prefectural level, it continued throughout the war in the village, where memory activists were heavily invested in promoting Oguri's legacy. For example, in April 1945, a month before the Oguri family sought refuge from the fire-bombing, the Oguri Artifact Preservation Society held a lecture "on the mourning of Oguri's death," which was attended by the Kurata Village mayor and members of the Gunma Prefecture Board of Education.[99] After the war, there were periodic memorial services, but the beginning of local and prefectural efforts to reinvigorate Oguri's image began in the 1950s. In 1953, the Gunma Prefecture Board of Education designated Oguri as "important to prefecture history," prompting the preservation society to change its name to the Oguri Kōzukenosuke Memorial Society.[100] The change from "preservation" to "memorial" (*kenshō*) is a non-trivial one. In this case, "memorial" in Japanese implies recognition and promotion, marking a subtle shift away from simple preservation.

The more productive connotation of "memorial" coincided with the law created in 1953 (*chōson gappei sokushin hō*) to encourage small towns and villages to amalgamate and form larger entities. After the formation of Kurabuchi Village in 1955, the head of the memorial society sent an official petition to the newly appointed mayor requesting that the village take over local Oguri commemoration and promotion, stating, "A new Oguri bust was erected in Yokosuka after the (San Francisco) Peace Conference thus making it an international memorial. His former bust now resides at the Tōzenji where his heroic form will continue forever. . . . [W]e should never forget that this was designated a Gunma Prefecture important cultural property and thus is a treasure of Japan. This acknowledges Oguri as an important benefactor of Japanese culture, and before long he will certainly be recognized as such by the Ministry of Education."[101]

Having failed in numerous attempts to gain national recognition for Oguri before the war, Oguri activists hoped that prefectural awards might lead to national honors. The Oguri Memorial Society was convinced that in order to promote Oguri on a larger scale, they needed to create a united front with the newly formed village office:

> Until today the Tōzenji parishioners have acted as the center of Oguri commemoration because the agents of murder, the Tōsandō, gave money to build his grave in Gonda. . . . Around 1935 the monk of the Fumon'in Temple, Abe Dōzan, started asserting that Oguri's severed head was located on his temple grounds. Many famous people such as Prime Minister Okada and Home Minister Tokonami Takejirō visited the grave, and even Tokugawa Iesato wrote the calligraphy for an Oguri monument erected at the Fumon'in. Until now

we have been waiting with bated breath to have our grave identified as the correct one. . . . [F]or these reasons we feel that the Tōzenji should not be handling
Oguri's commemoration (*kenshō*) alone. Events such as the donation of his bust
from Yokosuka and receiving prefecture recognition are meaningful for the
village, and will be good for the education and future development of culture in
the village. We would like to reform this group as a village organization.[102]

Although the Fumon'in had stopped holding its Oguri Festival, the
Tōzenji supporters feared losing ground against its prewar rival. Abe
Dōzan's research, as well as visits from important people, legitimized his
claim to represent Oguri's legacy. By connecting Oguri to local boosterism, Kurabuchi activists could project a unified, institutionalized front
to combat the rise of the Fumon'in or other potential rivals who might
appropriate Oguri's image.

From this moment, village officialdom dominated Oguri's promotion
because it portrayed the village to outsiders. The petition also used the
postwar keyword "culture," which John Dower explains was associated
in the immediate postwar period with the terms "peace" and "demo-
cratization," to distance Oguri's image from militarism.[103] More importantly, the association between culture and education suggested that
Oguri could help teach Japanese about their local and national identity
in a new Japan.

Throughout the 1960s, the memorial society increased its local presence
through the commemoration of several national historical events. In 1960,
Kurabuchi villagers celebrated the one hundredth anniversary of the signing of commercial treaties. This was a watershed year in the rise of Oguri
to historical significance due to his role in the 1860 embassy. They held
several lectures, a memorial service, and an Oguri Festival, the latter of
which had only been sporadically conducted since the end of WWII. This
pattern of lectures, memorial services, and festivals continued throughout the sixties and included participants from the Gunma Prefecture
Board of Education. The second memorial year occurred in 1965, the one
hundredth anniversary of the Yokosuka shipyard. Kurabuchi villagers
and representatives from the Oguri Memorial Society participated in the
Yokosuka anniversary, sending Oguri-related objects and documents,
much like their predecessors had done during the Yokosuka celebration
fifty years earlier. Locally, however, the memorial society decided to limit
the scale of the 1965 Oguri events in order to prepare for a far more significant event—the one hundredth anniversary of Oguri's death and the Meiji
Centennial.

The decision to restructure the Oguri Memorial Society from within
and widely promote Oguri's centennial marked a turning point in his local,

regional, and national appropriation. The society moved its office into the village hall, strengthening the bond between Oguri commemoration and village officialdom and accelerating the rate of memory production about Oguri. The society also created an Oguri research journal; conducted lectures; built an archive for Oguri items; held annual meetings; promoted Oguri's image on television, newspaper, and radio; erected new memorial plaques over Oguri's execution site and on Mount Kannon; and placed a monument where several of Oguri's retainers were buried.[104] The 110th anniversary of Oguri's death, held in 1978, attracted the attention of then prime minister Fukuda Takeo, who sent a message of congratulations. Born in Gunma, Fukuda praised Oguri's work as a financial expert and his efforts in Yokosuka, projects that contributed not just to the shogunate, he emphasized, but to Japan itself.[105]

Oguri's popularity originated within this context of *furusato* through a mutually beneficial relationship among memory activists, Gunma Prefecture administrators, and their counterparts in Yokosuka City. Like the Oguri family, Yokosuka also benefited from interactions with Gunma activists during the war. Toward the end of the war, the navy sent injured Yokosuka naval personnel to live in Kawaura and Mizunuma villages— both part of Oguri's memory landscape—where they manufactured charcoal to replenish the navy's depleted fuel sources.[106] In 1953, Yokosuka City donated its Oguri bust to Kurata Village, having replaced the old one with a new bronze. The bust was officially unveiled by Oguri Tadahito in 1954 at Tōzenji.[107] In 1956, Oguri's bust was joined by one of Kurimoto Joun, a copy of an original funded by former prime minister Inukai Tsuyoshi, who once criticized the posthumous court-rank system of the 1920s and 1930s. Kurimoto had no historical connection to the village, and initially seemed oddly placed at Tōzenji, but his bust strengthened the Oguri-related memory landscape connecting Yokosuka and Kurabuchi.[108]

The informal ties between Kurabuchi and Yokosuka finally blossomed into an official friendship city relationship in 1981. The announcement alluded to the historical connection between Yokosuka and Kurabuchi through Oguri and looked forward to "further cooperating in the future and conducting cultural exchange through Oguri's legacy."[109] "Cultural exchange" between the city and the countryside, in particular, plays on the *furusato* trope. Tourism from the city to the countryside was central to the official bond between Yokosuka and Kurabuchi. In 1981, nearly eighty officials from the Yokosuka city hall visited Kurabuchi in seven separate trips. The earliest group arrived in May by microbus, followed by almost monthly excursions until the official signing of the sister city agreement in December. These included several visits from Yokosuka

City Waterworks officials, who checked Kurabuchi's water quality; the deputy mayor, who examined the "quality of the autumn leaf landscape"; and members of Yokosuka's public relations department.[110] Finally, Kurabuchi announced that it would create a Yokosuka Citizen's Recreation Center. Kurabuchi built other sites to attract city people to its healthy, natural beauty, supported, ironically, by a rapacious dam-construction industry.[111] Together Kurabuchi and Yokosuka built Hamayū Sansō, a large mountain resort completed in 1987 at a total cost of 1.7 billion yen.[112] Yokosuka paid for the construction, and the village provided the land and also managed the business.[113]

Yokosuka was only one part of Oguri's expanding memory landscape. He became a martyr in Gonda/Kurabuchi, but his legacy and family's escape extended beyond the village. Since the postwar period, other parts of Japan began to identify with his legacy by contributing to his material commemoration. The Oguri archive built in the 1960s on the grounds of the Tōzenji includes objects donated by Gonda families, and many from people outside Gonda. Each narrates the Oguri story: documents, pictures, tools Oguri brought back from the United States, Western-style plates said to be a present from Oguri's wife when she traveled to Aizu, the palanquin that carried Oguri to his place of execution, a camellia descended from Oguri's original plant, and his Western-style chair, sold off by the imperial army and now owned by the mayor of a neighboring city.[114] Donated items further legitimized Kurabuchi and the Tōzenji's centrality in Oguri commemoration; those kept in private became public as they were integrated into the growing network of Oguri-related sites of memory.

Even the unseen objects—buried treasure—became part of this memory landscape. Much to the chagrin of memory activists, as his commemoration accelerated, the popularity of the buried-gold legend grew, becoming a major topic of inquiry for visitors to the Oguri archive.[115] Just as prewar memory activists such as Ninagawa and Toyohara continued their work after the war, with new activists joining the fold, so too did Kawahara and other prewar gold diggers. The war had interrupted the digging of the hole that Kawahara had begun in 1934, which had reached a depth of 248 feet. He received new financial support from a mysterious man from Nagoya who claimed to be a former retainer of the Tokugawa shogunate and one of Prime Minister Okada Keisuke's bodyguards during the attempted 2/26 coup (1936). In 1951, equipped with a generator, a water pump, an elevator, and thirteen workers, Kawahara restarted work on the hole. When asked about his progress, he responded, "Just as these Oguri documents show, I've already found a large stone, human bones, and ammunition, so the money is close."[116]

The 1950s Oguri gold rush seems to parallel the desperation that drove adventurers into the mountains during the 1930s. The same actors were involved: Kawahara in Yamanashi Prefecture, and the Mizuno father-son team on Mount Akagi in Gunma. These men were joined by a new treasure hunter, a former policeman named Saegusa Mosaburō. In 1949, Saegusa used the only ground radar detector available in Japan to search along the Mount Akagi plateau, and despite having found several promising locations, he still could not discover the hidden treasure trove. In 1956, on the verge of giving up, the Kobo Daishi appeared to him in a vision and encouraged Saegusa not to despair. Saegusa pressed on. "I'm ninety-nine percent there," he once told Hatakeyama Kiyoyuki, a treasure-hunting storyteller who visited Saegusa in 1959; he promised Hatakeyama that he would find the treasure soon.[117]

Of these would-be hunters, only Kawahara continued to involve himself directly with the Oguri memory activists and the Oguri family itself. In 1956, he appeared at the Tōzenji accompanied by an entourage driving expensive foreign-made cars; this included his new financial backer, whose wealth came from the pachinko industry. Kawahara told the monk to call together other local monks to hold a memorial service in Oguri's honor. He agreed to pay all of the expenses and left a small sum of money to show his sincerity. The monk called some thirty colleagues to the temple, but Kawahara never showed up, leaving Tōzenji with the bill.[118] Oguri's granddaughter Kikuko continued to denounce gold diggers, who she felt sullied her grandfather's good name. She blamed Kawahara's resurging celebrity on the news of a gold *koban* coin discovered in Tokyo during the postwar rebuilding of the neighborhood.[119]

Although memory activists and gold diggers competed for regional and national attention through Oguri, they represented parallel efforts to take advantage of the postwar rebuilding of Japan's commemorative infrastructure. The wartime bombing quite literally opened new ground for gold-digging projects, while postwar gold fever fit within the culture of desperation, fantasy, and hope that John Dower describes. Newly created villages, towns, and cities needed local identity more than ever, with money gradually finding its way into new identity-building activities. Local Oguri supporters received help not only from village and prefecture administrations but also from nationally recognized writers and essayists who offered their own interpretations of Oguri's life and death.

The subtitle for Kiya Takayasu's 1982 historical fiction about Oguri invokes the dual narratives of buried treasure and tragic hero: "killed for his buried gold and without having committed a crime."[120] By invoking Ninagawa's epitaph from the 1930s, Kiya clearly sympathizes with local activists, but the reference to buried treasure indicates that Oguri was, by the 1980s, also popularly known for the treasure legend. Kiya first encountered Oguri's story in the mid-1960s, during his tenure as the bureau chief for Jiji Press Ltd. in Gunma's capital, Maebashi.[121] He often visited Murakami Shōken, the Tōzenji monk, to conduct research about Oguri, and was in turn introduced to Ōtsubo Motoharu—the "Oguri authority," according to Kiya.[122] This establishes two points: that local boosterism in Gunma led Kiya to Kurabuchi, and that Kurabuchi supporters had already established appropriate "Oguri" experts beyond Gunma. Kiya faced immediate obstacles when he tried to publish his work. One reviewer, Tsunabuchi Kenjō, wrote, "I think it's very interesting that someone is trying to destroy Katsu Kaishū's image in contrast to the deification [he has undergone]. However, up to this point Katsu has always been portrayed as being a more important retainer than Oguri. I'm afraid that this is too complex an issue for readers."[123] The director of *The Birth of Tokyo* had shared the same sentiment years earlier: that the narrative pushed by memory activists was too extreme, too unlike the dominant Meiji Restoration story with which audiences were familiar. Here lies one of the major problems with Oguri's commemoration at the national level. Without a believable narrative, Oguri's story just will not capture a national audience; "Memories that deviate too much from convention are unlikely to be meaningful to large audiences or to be spread successfully."[124] Kiya rewrote his work, but his experience illuminates some of the difficulties of writing about Oguri in a popular culture milieu in which Katsu and others still enjoyed celebrity status.[125]

Conclusion

At the national level, Oguri's relative popularity did not survive the war. Other than Ninagawa, there were no longer any famous scholars, politicians, or journalists with a personal stake in rehabilitating Oguri. Before 1945, Oguri supporters elevated Oguri above the shogunate by depicting him as a military reformer, highlighting, for example, his procurement of funds for Yokosuka despite resistance from within the shogunate. But during the postwar period, Oguri became more representative of the conservative aspects of the shogunate. Even the director Ōsone rejected Ninagawa's anachronistic interpretation of Oguri's value to Japan and chose, instead, to use Oguri as a metaphor for loss at a time when Japan was coping with its own status as a loser. Ōsone also tried to depict the down-to-earth, human side of Oguri, just as Ii Naosuke had appeared in literature, kabuki, and film, but largely because of stylistic reasons, neither the film nor Oguri's portrayal in it became popular.

Significantly, Ibuse Masuji's short story was a retelling of local memory activity itself, one that shifted away from the Fumon'in to the Tōzenji and Gunma, reflecting the influence that local people have over the formation of a local hero's story. A new generation of local researchers, including Koitabashi Ryōhei and Ōtsubo Motoharu, continued the work of their predecessors. Yet with the demise of the history journal *Jōmō and the Jōmō People* after the war, they lacked a consistent venue for promoting Oguri locally. They filled this gap by creating a memorial society that quickly tied itself to Kurabuchi Village officialdom—the newly formed Kurabuchi needed a unifying identity, and memory activists needed more stable support. Help from the prefecture and growing interest in *furusato* throughout Japan led to more people and commemorative objects flowing into the village, and since at least the 1980s, Oguri commemoration expanded outward from the Kurabuchi center, creating an Oguri memory landscape that sometimes branched into other forms of commemoration unintended by memory activists, exemplified by the commercial relationship between Kurabuchi and Yokosuka. The memorial society and the Tōzenji monk have built plaques, memorial stones, incense altars, and statues throughout Gunma, Niigata, and Fukushima. These objects carved a museum into the memory landscape with a proper route and lessons to be learned. This became the foundation on which Oguri would become a national hero in Heisei Japan.

CHAPTER 5

Oguri and Japan's New Heroes during the "Lost Decade"

Every May on the anniversary of Oguri's execution, a festival is held on the grounds of the Tōzenji Temple. Each year, the festival seems to attract more people. The Tōzenji encourages farmers and local artists to set up tents and sell their wares. Throughout the day, junior high school students hold brief tea ceremonies. In the past decade or so, the festival has always included guest speakers: writers, local historians, university professors, manga artists, and other celebrities who participate in Oguri's commemoration. Sometimes Ogurin, the Yokosuka 150th *kaikoku* anniversary mascot, attends.

In 2005, the festival marking the 137th anniversary of Oguri's death included a separate event at a nearby gymnasium to accommodate larger crowds and featured a more structured program. Among the guests that day was Kimura Naomi, the manga artist who created the first Oguri manga. Murakami Taiken, the Tōzenji monk and current leader of the Oguri memory activists, was also on display, holding a brief ceremony in front of Oguri's and Kurimoto's busts, and the graves of fallen retainers. Later in the day, visitors listened to his lecture about Oguri while the Gunma Prefecture mandolin orchestra played in the background, meant to evoke emotion like a movie soundtrack. The temple itself was crowded with Oguri-related objects: watercolor paintings depicting his story; a model ship of the USS *Powhatan*, which carried the 1860 embassy to the United States (the escort ship, the *Kanrin Maru*, is purposely missing); posters of NHK's Oguri drama; Oguri-labeled sake; and Oguri *manju*. The object drawing much of the visitors' attention, however, was also the smallest. It

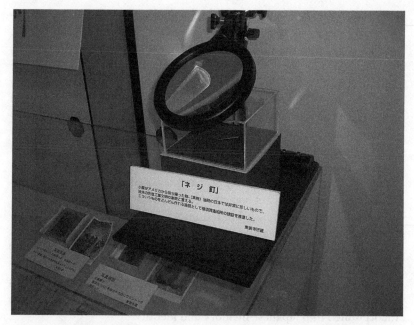

Fig. 3. Oguri's screw. The caption reads: "This is a screw Oguri brought back from America. This would have been a rare object in Japan at the time. It shows how advanced the West was as an industrial civilization." (Courtesy of Tōzenji)

was a lone screw, featured in the 2003 NHK drama as well as Oguri manga, like the one Oguri is said to have held on the return trip, while pondering how he would modernize Japan. Too small to be seen on its own, a large magnifying glass beckoned visitors.

Each of these objects might seem insignificant on its own, but together they highlight important lessons to be learned from Oguri's life and the twists and turns in his commemoration. These vehicles of memory invite us to, as Morris-Suzuki so eloquently states, "enter into an empathetic relationship with the people of the past [, which] in turn becomes the basis for rethinking or reaffirming our identity in the present."[1] This interaction between past and present explains why Oguri has received so much national attention during the Heisei era (1989–present), when there has been a veritable Oguri boom, itself part of a national *bakumatsu* boom. Since the "lost decade" began, everyone from political pundits to local memory activists has invoked the late Tokugawa period as a model for solving Japan's contemporary problems. Restoration losers have risen to the forefront because they represent an alternative to the historical path forged by the Meiji government, which is often seen as the source of Japan's current problems. This is reflected in historical novels, manga, television programs, and films, where Oguri, Ii Naosuke, the Shinsengumi, and Aizu samurai have resurfaced as new heroes.

In addition to their value in critiquing the Meiji legacy, these newly defined heroes are also part of two major memory trends of the Heisei era. Neither memory trend is new, but each has taken on a new significance during the 1990s, when Japan's putatively secure economic, political, and social legacies have eroded—in perception if not always in reality.[2] The first is growing global interest in WWII memory and commemoration, which accelerated during the 1990s and included participation at various levels of society: studied by academics, discussed by public intellectuals, displayed in museums, and featured in popular culture. This has been especially intense in Japan, where wartime atrocities remain unsatisfactorily addressed domestically and continue to offend Pacific neighbors. During this new "age of memory," as Narita Ryūichi has called the 1990s, the "us" in "our memory" has been called into doubt as new historical voices, such as those of the comfort women, have been incorporated into Japanese memory and identity.[3]

The second memory trend, nostalgia for the early modern period, seems unrelated to the first. Carol Gluck has summarized the different Edo memory tropes of the 1990s: optimistic versions of an Edo period from which all things uniquely Japanese emerged, already modern, and to which the Japanese identity should return.[4] Here I am interested in a subset of Edo memory, the Meiji Restoration years, which have been compared to the Heisei period by pundits, writers, and film producers; even politicians have gotten into the act—quite literally. Every few years, politicians from opposing factions join company presidents and journalists to perform in an ongoing theater production titled *The Heisei and Bakumatsu World Renewal* (*Heisei bakumatsu yonaoshi geki*).

This strain of Restoration memory tends to disavow the Meiji legacy. One argument, for example, states that whereas the Meiji government fired the daimyo and dissolved the domains, Japan today should set up daimyo and let regional governments take the lead in transforming Japan.[5] A strong, centralized, bureaucratic state brought Japan into modernity and bestowed it with a robust economy, but the Meiji bureaucratic legacy has worn out its use. Others argue that "Meiji" is synonymous with Japan itself and call for "casting off the Meiji regime" during the Restoration bicentennial (2018) by pronouncing Japan a failed country.[6] Japan, as a project of the Meiji legacy, is lost, but this loss paves the way for a new beginning.

War Memories: Rectifying Heisei Japan through the Boshin War

The war-memory anxieties that occupy discourse at the national level also play out at the local level, but they are channeled not through WWII but through the Boshin War of 1868. Although the Meiji Restoration is often portrayed as a relatively bloodless event compared to modern revolutions elsewhere, thousands died during the Boshin War in the northeast, particularly in the former Aizu domain. Thus, there remains an alternative narrative of "the war," as it is known there, similar to the confederate identity that still thrives in parts of the American South. The demand for apology, and the offer of reconciliation by Aizu's former enemies, allow local citizens to participate in the nationwide apology discourse originating from WWII memory. Although the situation disallows Oguri memory activists to demand an apology, they seek rectification for past injustices by staking a place for him within the national consciousness.

Participation in this national-level discourse is typically unavailable to small communities where localized experiences of WWII are often appropriated into a larger Japanese whole—local Hiroshima and Nagasaki suffering, for example, has been subordinated by the state to represent national victimization. Moreover, the ubiquity of local WWII sites of memory, including museums and document centers (*shiryōkan*), erases their uniqueness—every town has one, but not every town can boast historical connections to the Boshin War. Through Boshin commemoration, local groups can re-create the same discussions held at the national level regarding memory and apologetics: the relationship between victims and victimizers, whether or not to apologize for the past aggressions, and using the past to address problems of the present.

The two parties most involved in Boshin memory are the former Aizu domain, now Aizu-Wakamatsu City in Fukushima Prefecture, and Chōshū, typically represented by Hagi City in Yamaguchi Prefecture. Residents in Aizu-Wakamatsu, where the Boshin War took place, feel victimized by the Chōshū-led forces that killed, and left unburied, some three thousand of their ancestors, sending the survivors, labeled as "enemies of the court," into neighboring domains as refugees. In a reversal of the WWII victim and victimizer relationship, it has been the putative victimizer—people from Yamaguchi—who have tried to reach out to the victim first. Hagi City officials tried to establish sister-city relations with Aizu-Wakamatsu during several commemorative events, including the Restoration centennial (1968) and both the 120th and 121st anniversaries of the Boshin War (1987 and 1988), but Aizu officials rejected all such proposals. In 2007, when Prime Minister Abe Shinzō, a native of Yamaguchi, campaigned on behalf of local Liberal Democratic Party (LDP) candidates in Aizu-Wakamatsu, he began his speech by apologizing for his ancestors who "caused trouble" in the area.[7]

Private citizens initiated the Aizu-Chōshū reconciliation. In the fall of 1996, a local Aizu-Wakamatsu citizens' group put on a play about a woman from Aizu who falls in love with a young man from Hagi and, despite opposition from her father, eventually marries him. Without consulting city officials, the group invited the Hagi mayor to attend the performance in Aizu-Wakamatsu; he agreed to attend as a private citizen.[8] Invoking "private" status is a strategy more famously adopted by prime ministers who wish to avoid criticism when visiting Yasukuni Shrine to commemorate the war dead on August 15. In this case, Aizu-Wakamatsu City officials did not know about the invitation until ten days before the event. No mayor from either city had ever traveled to the other, nor even met; such an occasion would be historic. The unofficial visit by the Hagi mayor did not receive a warm reception. Caught off guard, Aizu-Wakamatsu City officials contacted the Hagi mayor's office to make it known that they wished him to visit as a private citizen, not as a public figure.[9] When the Hagi mayor arrived by train in the morning, no city official from Aizu-Wakamatsu was there to meet him.[10] Both mayors attended the play, but they sat far away from each other and did not share a greeting until after the play, and then only briefly, not even shaking hands; one journalist described it as an awkward encounter.[11] In the summer of the following year, a similar citizens' group in Hagi responded by staging a musical version of the play in both Hagi and Aizu-Wakamatsu. These performances were cited as an important first step toward more official interaction between the two cities.

Historical reconciliation led to thoughts about mutual city development: native goods from both cities were displayed in Aizu-Wakamatsu, artisans started researching how to mix Hagi and Aizu glass-making styles, and there were talks about how to encourage tourism between the two areas. High school educators, concerned with a poll showing that 32 percent of Aizu-Wakamatsu citizens had bad feelings toward Hagi, started a pen-pal program and gave team-taught lectures on the theme of reconciliation between the two areas. Hagi City even hosted a team kendo tournament involving Hagi High School, Aizu High School, and Konan High School from Kagoshima Prefecture, the former Satsuma domain that was a one-time rival to both Chōshū and Aizu. Unfortunately for Aizu-Wakamatsu, Aizu High came in last place.[12]

Private memory activists pushed more difficult topics onto the memory landscape by holding a conference with scholars from the former Aizu, Chōshū, and Satsuma domains. In 1998, scholars, writers, and city officials participated in a 130th anniversary conference of the Boshin War in Kakunodate, Akita Prefecture. Previously, except for a fiftieth anniversary in Iwate Prefecture, no major event by the losing side had been held during an anniversary of the Boshin War. The Kakunodate citizen's history group that organized the conference wanted to bring attention to Akita's victimization during the Boshin War, feeling, no doubt, that Aizu-Wakamatsu received most of such attention. One member commented that only Akita Prefecture had been doing well enough to celebrate its status as a loser, especially since other prefectures in the northeast had been having trouble in recent years.[13] But the timing seemed out of place. Most major anniversaries occur during intervals of fifty or one hundred years, such as the 150th anniversary of *kaikoku* held in Yokohama and Yokosuka in 2003, or the 150th anniversary of Japan–U.S. relations held in 2010 in New York and elsewhere.

In this case, the timing coincided with a major theme of the conference: reflecting on the past during a chaotic historical moment. Only ten years earlier, in 1988, Japan was still benefiting from positive economic growth, but by 1998, not only had the economy yet to recover but natural disaster (the Hanshin earthquake in 1995), religious terrorism (Aum's sarin gas attack in 1995), and numerous government scandals battered Japan's self-confidence from every direction. Many conference speakers blamed the post–Boshin War government: its centralization, militarization, and, in the words of one panelist, "the idea that bureaucrats will take care of everything, and anyone who says otherwise is in the way, an idea that is now disappearing."[14] Others connected Boshin commemoration to recent WWII memory. Hoshi Ryōichi, a renowned historical novelist, commentator, and Fukushima native, emphasized this point: "Satchō's massacre of 3000 people outside of Aizu Castle became the model for Japan's invasion of Asia."[15]

The conference ended with no clear resolution between the two sides of the Boshin War. On the Aizu side, memory activists continue to feel that they have been wronged on three issues: the disproportionately large amount of land, nearly 87 percent, taken away from the Aizu domain relative to other losing domains; Aizu refugees being forcibly moved from their homes; and the mistreatment of Aizu dead around its castle. From the Aizu perspective, Chōshū deserves special blame because it sought revenge for Aizu's alliance with Satsuma in a Tokugawa punitive mission against Chōshū.[16] The Satchō side is unable to address these complaints materially. Unlike the victim versus victimizer relationship concerning comfort women, reparations are not an issue, nor could they be, for the same reasons that complicate demands for an apology: domains no longer exist. A speaker from Kagoshima expressed this succinctly. While acknowledging the suffering caused during the Boshin War, he argued, "I'm not sure if Satsuma has to apologize, especially since there is no longer any agent (*shutai*) to apologize."[17] Another Satchō defender adopted the familiar rhetoric used to defend Japanese responsibility during WWII: the victimizer was also a victim. Chōshū was on the winning side, but Chōshū commoners suffered. They were forced to fight as part of an irregular unit, called the "kiheitai." Its founder, Takasugi Shinsaku, although heroized nationally, was not so well regarded locally: "his niece had a terrible time trying to find a husband, and when she finally married, it was as a second wife."[18] Even though Chōshū identity today might include being a "winner" in the Restoration, that narrative silences the great majority of commoners who gained little from the victory.[19]

Another obstacle to the Aizu-Chōshū reconciliation concerns the multifaceted nature of memory-related identity. Aizu and Chōshū maintain narratives about their relationship to each other, which can presumably be repaired through a combination of the sort of dialogue and cultural exchange previously described. But Chōshū also has a relationship with Satsuma that is a valued part of Chōshū identity. The Satsuma-Chōshū relationship, exemplified in the seemingly innocent high school kendo competition, complicates reconciliation between Aizu and Chōshū; the former connection was based on a shared animosity toward the shogunate and, by extension, Aizu. However, when Shimonoseki (Chōshū) and Kagoshima (Satsuma) became sister cities in 2004, the announcement celebrated that history. As the Shimonoseki mayor gushed, "It felt just like *bakumatsu* times, joining together, and attacking Tokyo in order to reform Japan."[20]

In the end, the forum provided no solution for inheritors of Chōshū's legacy who might be sympathetic to Aizu's demands for apology.[21] Hagi and Aizu-Wakamatsu cities have been willing to repair their relationship by promoting tourism and sales of local products, while Aizu residents themselves, those few who care about the issue at all, are not sure what reconciliation should look like, even though some believe that it is needed. Still others do not want reconciliation because being the "loser" is critical to Aizu identity. One speaker at the conference, a professor from Kagoshima University, encountered such a reaction while having drinks late one night during the conference. An Aizu local told him, "Let's [former Aizu and Satchō] become better friends, but we will not reconcile."[22] Perhaps this is instructive in understanding the winner-loser dichotomy in Asian memories of WWII: China and South Korea might draw nearer to Japan but will not forget Japan's violent past. In this way, the Aizu historical memory is unlike that of WWII victims, who still bear the scars from their experience; the Boshin War is safely distant from the present but recent enough to influence a modern regional identity. Both historical memories are connected through the identity crises exacerbated by the 1990s and the shifting relationships between the Japanese state and small communities—whether those communities are domestic or foreign ones—where people demand that the dead and scarred not be trampled under the narratives of a triumphant Meiji, or postwar, Japan.

Lost Decade, Lost Gold

Interest in Oguri's buried treasure resurfaced during the 1990s for the same reasons it had during the 1930s and 1950s, but this time the boom grew in proportion to the economic crisis. As personal fortunes became the first victims of the lost decade, Oguri's national reputation became synonymous with the buried-gold legend. The Kawahara men continued to dig for gold in Yamanashi Prefecture, as did the Mizuno family on Mount Akagi. The gold legend parallels Oguri commemoration: just as the Tōzenji eclipsed the Fumon'in as the new headquarters for memory activism, the Mizuno heirs overtook the Kawahara as the new gold-digging gurus. Mizuno Tomoyuki became the new buried-treasure "sensei" for would-be treasure hunters seeking to offer their own treasure theories.

Even people unfamiliar with Oguri's name were likely to have heard of the treasure legend. The Tokyo Broadcasting System (TBS) featured a series of digs on Mount Akagi from 1990 to 1996 in its appropriately named variety show *Gimmie a Break* (*Gimia bureiku*). The project grew in popularity throughout the 1990s, peaking in 1993 and 1994 when it captured a staggering 20 percent of the viewing audience.[23] In each sequel, TBS introduced a new element to the dig: two psychics from the United States; an esoteric priest who used his powers to help guide the dig (and, in one episode, exorcised the spirit of a dead treasure laborer); radar equipment; a helicopter-carried sensor; and, in the final episode in 1999, an industrial boring drill, which indicated that only solid rock was underneath their two-hundred-foot-deep hole.

The TBS treasure hunt encouraged a decade of treasure hunters to go searching in Gunma. Kawaguchi Sunao, who has kept track of the phenomenon, notes that every ten to twenty years there have been treasure-hunting booms, but the Tokugawa-related treasure boom of the mid-1990s was broader and lasted longer than any previous boom. "In 1994 and 1995 many people asked bus drivers outside of the Takasaki and Maebashi train stations for directions [to appropriate locations] and announced their treasure-hunting intentions," noted Kawaguchi, "and more people visited the prefecture library looking for maps and Oguri-related materials [to help them find treasure]."[24] Subsequently, more excavation sites not associated with TBS appeared on television. For example, Fukasawa Kōji, an elderly man in Seta County, Gunma, continues to dig what has become a ninety-eight-foot-deep hole on his property. Fukasawa began searching in earnest in 1973, remembering the legends his grandfather had told him. When asked by a television documentary crew what he would do with the money, he responded, "I haven't really thought about anything like that; I've just been focusing on finding it. . . . [I]t's buried just one more meter below."[25] This is a familiar refrain among Tokugawa treasure hunters; Mizuno Tomoyuki has often stated that "90% of the Tokugawa buried treasure mystery has been revealed; after the treasure has been discovered, the other 10% will fall into place."[26]

In his history of treasure legends, Crossan notes that because the treasure is never found, the story itself never ends; it remains as long as people continue to search or as long as the tale is remembered.[27] It would be easy to dismiss all treasure hunters as delusional at best, or as scam artists at worst, but Mizuno makes similar claims to memory as do Oguri memory activists. He confidently denounces treasure hunters for lacking historical evidence to back their claims, and he even questions his own family's historical vigor: "A lot of what my grandfather said as 'it *appears* that' changed to 'it *was* that.'"[28] Only he can find the treasure because only he possesses his grandfather's research from the early Meiji period. These documents have become symbols of legitimacy, allowing him the greatest claim to the legend's legacy. Since the 1970s, Mizuno has published several books about his family's story, their search for the gold, and, in his later works, his life philosophy as embodied in his search. He has even become the subject of an independent documentary.[29] He does not care about the money, which, he notes, would legally become the property of the state: "Once I find the gold I can die; I will have nothing to do with it after that."[30] Instead, he digs for history, to uncover the history of losers—like Kodama Sōbei, who served the Tokugawa shogunate (whose treasure notes passed to Mizuno's grandfather) and who was buried without receiving due recognition.[31]

The TBS special and the subsequent boom in treasure hunting in Gunma brought Oguri into the national limelight as never before. The first books published in the early 1990s with Oguri in the title are all about buried treasure. The first, Gōrō Tenkyū's 1991 *Oguri Kōzukenosuke's Treasure* (*Oguri Kōzukenosuke no hihō*), is a mystery novel set in the early Meiji period that has little to do with Oguri, although it uses Ninagawa's biography as a reference. The second, published in Gunma in 1992, is a mix of local Oguri history that sympathizes with local portrayals of him, but it quickly turns into an argument about the gold's true location in Kuragano, where the author is head of a local archive, not on Mount Akagi.[32] The current monk of the Tōzenji, Murakami Taiken, wrote of this work, "It's just another treasure hunting book and an attempt to bring attention to Kuragano—a mess of fact and fiction."[33] Even after TBS ceased its gold digging in 1999, NHK documentaries about Oguri used the treasure legend as a hook to draw in the audience. Such was the case in the 2000 serialized documentary *You Thought You Knew* (*Shitteiru tsu-mori?*), which featured Oguri. Celebrity guests split into treasure believers or nonbelievers, and offered commentary in between reenactments from Oguri's life. In 2008, TBS revived its defunct show *Gimme a Break* by conducting one last dig. Producers brought along a much-older Mizuno, still weak from lung cancer surgery, and a new host, the popular SMAP member Katori Shingo, but it failed to inspire audiences as it did during the 1990s.

Since the end of the post-bubble recession, the Tokugawa legend has faded into the background, though it still draws the occasional tourist to the Tōzenji. The most zealous Oguri supporters face a precarious situation; they may deny and despise the legend, but Oguri's memory probably cannot exist without it. Local historians quickly disavow the story but use it as a tool for arguing why Oguri should be remembered. This makes for a complex relationship between memory activists who reject the legend yet benefit from the exposure, and treasure hunters who might sympathize with Oguri's commemoration yet know their activity offends.

A Decade Lost, a Hero Found

The gold legend alone does not account for the Heisei Oguri boom. In-stead, as with other Restoration losers, Oguri's story fit within the crisis of the "lost decade." From the end of WWII until 1990, only five books contained Oguri in the title. Besides the two mentioned previously, one was another biography by Ninagawa written shortly before his death, and the other two were by a local researcher in Gunma Prefecture. Be-tween 1990 and 2010, twenty-two books had been published, not includ-ing manga or republications of earlier Heisei period works. Many of

these were authored by, or received direct input from, Oguri research-
ers and memory activists in Gunma. His popularity, in turn, helped
Gunma officials petition NHK to feature Oguri in its annual New
Year's historical drama, part of the 2001 Oguri Campaign held in
Gunma and Yokosuka.

This reappropriation of Oguri sets him apart from more familiar model
heroes. For example, some pundits ask who will be the Saigō Takamori
or Sakamoto Ryōma to introduce, unexpectedly, a new energy into these
troubled times, as calls for a Heisei Restoration mimic similar demands
for Taishō and Shōwa Restorations. The concept of a "Heisei Sakamoto
Ryōma" even became the focus of competition between two candidates
for the governorship of Kōchi Prefecture, Sakamoto's birthplace, and
both attended Sakamoto-related events.[34] Others either argue that Japanese should not be looking for the next hero but the next political system,
or caution that looking for a new hero might lead to bandwagon slogans
and reforms that could be too severe, as they were during the Meiji Restoration.[35] In the latter case, the author alludes to the violence perpetrated
by the Restoration victors, arguing that the two major terrorist events in
Japan were the assassination of Ii Naosuke and the 2/26 incident of 1936,
both of which sent shocks through the leadership.[36]

The most influential rewriting of Oguri came from two books released
during the late 1980s. The first, Sakamoto Fujiyoshi's 1987 *The Life of
Oguri Kōzukenosuke* (*Oguri Kōzukenosuke no shōgai*) was the first thorough biography since Ninagawa's and certainly more academic. Sakamoto cites his sources and critically reads other Oguri biographies. The
biography continues many themes addressed in his earlier *Bakumatsu
Restoration Economists* (*Bakumatsu ishin keizaijin*), in which he credits
Oguri with introducing to Japan both modern management, mainly as
the Yokosuka project manager, and modern accounting.[37] As an economic
historian writing in the midst of the bubble economy, it is no surprise
that Sakamoto emphasized Oguri's role as a father of Japanese management, Japan's economy, and Japanese capitalism. He uses the Hyōgo
Shōsha as an example of true capitalism in contrast to contemporary
Japanese capitalism. He argues that small and medium-sized companies
are unable to trade their stock because large financial institutions cover
their risks; they become, in effect, too small to fail.[38] He acknowledges
the assistance of the Tōzenji in his research, and cites the local Oguri
journal, *Tatsunami*, but is quick to note that although his book is a re-
quiem to Oguri, "[his] judgment is based on objective observation, not
upon any sympathy toward the losers (*hōgan biiki*)."[39] Sakamoto replaced

Ninagawa as the most referenced Oguri biographer in novels, television programs about Oguri, and manga.

Japan's most popular historian, Shiba Ryōtarō, offered the most visible reassessment of Oguri. In *A State Called "Meiji"* (*Meiji to iu kokka*), Shiba calls Oguri one of the fathers of Meiji Japan and states, "Oguri was a patriot in every bone of his body, but he wasn't the type to talk about patriotism. Real patriotism is not about getting loaded and letting the tears flow while talking big. In such times, there are as many of those kinds of patriots as there are dogs in the mountains, fields, and towns barking so loudly it bursts my eardrums. Oguri was not that kind of patriot. He sent a new energy through the day to day affairs."[40] Oguri became a model of how a true bureaucrat should operate: without concern for shallow political speech. The entire second chapter, "A Legacy from the Tokugawa State," follows the typical themes, and in the same order, long espoused by memory activists: Oguri's dislike of extravagance (for example, eschewing poetic excess), his frankness, his rivalry with Katsu Kaishū (and why Oguri is the better man), his fruitful relationship with the French, the creation of Yokosuka for Japan (not just the shogunate), and the Meiji government's silencing of Oguri's successes. Narita Ryūichi has explained that Shiba's changing assessment of Oguri from *There Goes Ryōma* to *A State Called "Meiji"* is more than a simple shift in temporal context between the 1960s and 1990s. Narita argues that Shiba thought carefully about how he had presented history over the years, and thus it was not the case that changing contexts necessarily led to reevaluations of the past. "Shiba's reassessment of Oguri, the complete opposite of his previous one, demonstrates how drastically the meaning of "Japan" and "postwar" has changed over the twenty-three years separating both works."[41] Narita does not offer examples of how society had changed, but when Shiba published his essay in 1989, Japan's Nikkei had peaked, and Oguri and other Restoration losers had yet to become models for a lost Japan. Moreover, most of the Oguri discourse tied back to the local people who had been promoting Oguri with greater acceleration since the 1980s.

Since these two publications, Oguri's popularity has risen alongside the growing Restoration memory. As the focus of the lost decade shifted from economic woes to critiques of politicians and bureaucrats, so too did Oguri move from being a paragon of Japanese business, as emphasized in Sakamoto's biography, to a model politician for the twenty-first century. Takahashi Yoshio's *Japan in Trouble: Oguri Kōzukenosuke and Minomura Rizaemon* (*Nippon taihen: Oguri Kōzukenosuke to Minomura*

Rizaemon) fits the first category, defined as an "economic novel" that should be a guide for a Japan suffering from a burst economic bubble and failing banks.[42] Likewise, Oguri and Tateishi "Tommy" Onojirō are portrayed as new heroes for an economically distraught Japan in Akatsuka Yukio's *Have You Heard the Tommy Polka: The Bakumatsu for Oguri Kōzukenosuke and Tateishi Onojirō* (*Kimi wa Tomii poruka wo kiitaka: Oguri Kōzukenosuke to Tateishi Onojirō no "bakumatsu"*). For much of its history, the author argues, Japan has been an industrial country, but Oguri and Tommy embody the latent potential Japanese have to participate in the increasingly important global information society. They are agents of dispersion, the essential feature in an information society, rather than the tools for centralization characterized by the Meiji leadership.[43] In that same year (1999), Arai Kimio, president of Tokyo Group Inc., declared Oguri to be the sole model for a time when Japan faced the downfall of safe banks, disappearance of lifetime employment, and decreasing land value.[44] Since 2001, other writers have emphasized Oguri as a father of Japanese modernization and nation building.[45]

Depictions of Oguri as a model economist and politician are not new; they have been part of his rehabilitation since the Meiji period. What *was* new during the 1990s was how these themes were woven into his relationship with commoners, particularly Gonda peasants. Adding this element to his story highlighted his role as a Japanese first and Tokugawa retainer second. This became particularly important, as bureaucrats of both the shogunate and Heisei Japan were seen as the root problem within the government. As Hayami Akira observed, "Half of the *bushi* received salary but had nothing to do, and yet they could not be fired, just like bureaucrats [today]."[46] Blame for poor rural economic development, eroding rural populations, and corrupt local bureaucracy has often been attributed to the adverse influence of the central government. Oguri rose above petty status distinctions that characterized the time period, but his potential as an impartial leader, evidenced by his close ties to peasants, was dashed by his execution in Gonda. The moral of Heisei Oguri stories is that the countryside is innocent; bad things happen there only when the central government intervenes.

The earliest and most nationally recognized historical fiction novelists to write about Oguri also participated in Gunma-centered memory activism. The first Heisei-era Oguri novel, *A Sudden Rain in Jōshū Gonda Village: The Life of Oguri Kōzukenosuke* (*Jōshū Gonda-mura no shūu*), by Hoshi Ryōichi, departs from the typical narrative, which follows Oguri's service to the shogunate, and focuses entirely on his struggle in Gonda. Hoshi, himself a native of Fukushima and a longtime writer about the

Restoration in Aizu, connects Oguri's personal story to Aizu, where Oguri's family took refuge and his retainers died fighting, tying Oguri's story into the "loser" identity fostered there. Hoshi is not a dispassionate chronicler of the Boshin War but actively participates in the memory landscape by attending, for example, such events as the Kakunodate conference.

Dōmon Fuyuji, also a well-known historical novelist, first serialized an Oguri story in the Gunma and Kanagawa prefecture newspapers in 2001, the peak year of Oguri commemoration in Gunma.[47] Although Dōmon tries to reach the widest possible audience by covering all aspects of Oguri's life, he has introduced more of the Gunma side with each revision. As one critic noted, early versions of Dōmon's novel read like a business book, advocating Oguri as a new business hero and gradually portraying Oguri as a broad reformer. Only later versions became true historical tales (*monogatari*) by featuring Oguri's private side, the plight of his family, and his interaction with his fief.[48]

Hoshi and Dōmon have bridged the gap between a local audience for Oguri memory activism and a national one. Beyond simply fictionalizing Oguri's story, they have appeared as commentators in Oguri-related television programs. Hoshi appeared on the TBS gold-digging program to disavow the legend, while Dōmon was employed as a guest commentator on Oguri-related programs on NHK. Dōmon was an invited speaker at the 2001 Gunma Symposium, and both men have spoken at Oguri Festivals in Kurabuchi, which no doubt drew bigger crowds. Hoshi and Dōmon have become the newest resources in Oguri commemoration, building on the textually mediated memory about him first established in the Meiji period, but updated and expanded for a Heisei readership.

Dear, Have You Quit Again?

In 1999, the mayors of Kurabuchi Village and Yokosuka City and the Gunma Prefecture governor visited NHK's president in Tokyo to ask him to consider featuring Oguri in a Taiga drama—an annual, yearlong historical program. Governor Kodera told the local press, "History is always written by politicians so a proper evaluation never occurs. Oguri is a textbook case of this. Because he was on the shogunate side, his reputation is pretty low. I'd like to see a historical re-evaluation."[49] Of course, more than a few local government officials saw the obvious benefit to be accrued from national television exposure; one study has shown that the Taiga drama series leads to significant tourism profit for locations heavily featured in the series.[50] Although a Taiga drama never materialized, the lobbying effort was successful; NHK finally agreed to feature Oguri in the annual New Year's historical drama, *Dear, Have You Quit Again?* (*Matamo yametaka teishūdono*), which aired in January 2003. As a prelude to the historical drama, the network introduced him to audiences in November 2002 through the educational program *That's When History Moved* (*Sono toki rekishi ga ugoita*). The short, thirty-minute program featured commentary by Dōmon Fuyuji. The subtitle, *Modern Japan Started from One Screw* (*ippon no neji kara Nihon no kindai ga hajimatta*), invokes the story first advocated by Murakami Taiken—Oguri wanted to change Japan from a country of wood to one of iron. In fact, NHK staff often visited the Tōzenji to research Oguri. It is not known how many viewers watched either program, but it reached the widest possible audience with the biggest impact. The process of achieving this goal was as important as the result—it was the first time that village, prefectural, and Yokosuka City officials had joined forces to commemorate Oguri at the national level since erecting his bust in Yokosuka in 1922.

The degree and range of Oguri's appearance in Gunma and Yokosuka eclipsed all earlier commemorations of him. In both areas, Oguri was celebrated through conferences, festivals, tours, lectures, newspaper and magazine articles, and an ever-growing number of plaques and other memorabilia. Major symposiums were held in Maebaeshi City, the capital of Gunma, and Yokosuka in the fall of 2001. Dōmon Fuyuji, the Yokosuka mayor, and the governor of Gunma attended both events as speakers; the Yokosuka symposium also included two professors, one from France and one from Yokohama, to add an international and academic voice to Oguri's memory landscape.[51] Some in the audience remained skeptical about the newfound zeal for Oguri. When asked about the project to promote Oguri as a topic for a Taiga drama series, Governor Kodera remarked: "The series is not just about regional boosterism. I'm not hoping for a drama [that focuses just on] this locale. I want them to make a story about someone who tries to have faith in himself during a time of chaos and leaves something for future generations. . . . [I]t's not just about Yokosuka or Gunma. If we really examine the historical evidence there will probably be things that we would not want known, but I don't mind showing those either."[52]

It is not surprising that Kodera's comments connect Oguri's story to the familiar trope of comparing the Heisei crisis to the Meiji Restoration; more revealing, however, is his allusion to the possible negative repercussions of Oguri's history. He did not elaborate, but the violence experienced in the Gonda area due to Oguri's presence there was still an issue even during the Oguri campaign. For example, in 2001, the Oguri Research Society in

Kurabuchi planned to build a monument to those who died during the attack on Oguri in 1868. Although it was supposed to be unveiled at the 135th Oguri Festival, as of yet, no such monument exists.[53]

The Taiga drama series can also be a source of historical tension, especially when dealing with controversial historical topics. Most Taiga dramas are set in the safely distant past. Only a few series touch on the twentieth century, and only a handful address the Restoration years. Restoration dramas typically follow trends already established by other forms of popular culture, mainly literature. For example, *There Goes Ryōma* (1968) appeared shortly after Shiba Ryōtarō's best seller by the same name. The more recent *Shinsengumi!* (2004) was not based on a single best-selling novel but on the group's reappearance during the 1990s in manga, film, and historical fiction. The producer tried to tap into the Shinsengumi's popularity among young people, especially young women. "Young people in their twenties can relate to this version of the Shinsengumi," said scriptwriter Mitani Kōki. "In those days they couldn't trust anyone so they followed their own minds. Young people during the *bakumatsu* were frustrated, but through trial and error they took their chance in Kyoto. I want to show young people [the Shinsengumi] eagerness."[54] NHK, a public entity, tends to avoid controversial history topics, and for good reason. During a review of NHK's budget in 2004, Matsuoka Masuo, the Upper House representative for Yamaguchi Prefecture, attacked NHK's portrayal of the Shinsengumi, which he called a "terrorist group," labeling the program a "variety show" rather than a historical drama.[55] Another Diet member agreed, adding that heroizing or legitimizing the Shinsengumi "goes against historical fact" and should be avoided.[56]

The demand to account for historical accuracy, the core issue among Oguri activists, was adopted by the producers of his NHK drama. According to the creators of the show, "Each historical drama is supposed to make one think. In this one we wanted people to think about the issue 'might makes right' (*kateba kangun*)." They believed that Oguri's history embodied an alternative narrative to the dominant version of Meiji Restoration: "One will find historical fact that is not taught in school. Victors in war twist history to their own liking and hide that which does not help them."[57] The producers took liberties with the historical facts of Oguri's life to emphasize Oguri's importance. For example, in the movie, when Oguri muses about the Yokosuka shipyard falling into the hands of the shogunate's enemies (it is not referred to as an iron foundry in the movie, though it was at the time), he utters to Katsu, not Kurimoto Joun, the famous line "selling a house with a storehouse attached." Viewers were likely not familiar with Kurimoto, but Katsu is a standard figure taught in most high-school level history courses. Upon hearing Oguri's line, Katsu is speechless. Finally, he says to Oguri, "I thought you were just concerned about the shogunate, but now I know that's wrong." Katsu's reaction has more impact; throughout the film he criticizes Oguri for being a stubborn supporter of the shogunate who is unable to see the greater good for Japan in capitulating to the court. This historical interpretation, long advocated by memory activists, is powerfully relayed to the viewer—the idea that even Oguri's rival, the Restoration hero Katsu, acknowledges Oguri's contributions.

Oguri and Katsu are also juxtaposed as two types of modern men: the faithful family man versus the cheating husband. Vitriol against Katsu among Oguri activists has been as passionate as their love for Oguri, a view shared by Dōmon and Hoshi.[58] Thus, unlike *The Birth of Tokyo*, in which Katsu is a marginal character whose relationship with Oguri is never problematized, he is a central figure in this drama: an anti-hero in the political story, and a villain in the human one. The relationship between Oguri and his wife, Michiko, is meant to represent a modern, slightly Western one: they go on dates together, hold hands during walks, and ride on the same horse, and in one of the last scenes, he tells her in English, "I love you." Whereas Michiko appears young and innocent, Katsu's wife is old and tired, surrounded by kids. During one meeting between the two women, Katsu's wife lets it slip that she longs for her geisha days, when she did not have to work as hard. On her way home, Michiko runs into Katsu and catches him kissing a young woman behind a tree along the side of the road. Katsu sees her pretend to ignore him, and he calls out to her. She shouts, "Is this any way for [a newly promoted] navy commissioner to behave?" Grabbing her violently and pulling her close, he answers, "Even though people are of different birth, they all *do* the same thing." "Vile thing," she retorts, and he admits with a smile, "There are just some things I can't give up." In an age of growing political scandal, Katsu represents the worst in even the most able politicians, while Oguri remains pure in his work and personal life.

There is no available data concerning viewership of the NHK drama, but evidence suggests that it has helped promote Oguri nationally. In 2008, Minister of Justice Hatoyama Kunio discussed Oguri's drama at a press conference following a cabinet meeting. When asked about his pre-meeting conversation with Prime Minister Asō Tarō, Hatoyama responded, "I told him about this drama aired quite a while back about the Tokugawa Finance commissioner Oguri Kōzukenosuke, and how he went to Gunma to pursue agriculture only to be killed by the imperial army over buried treasure. I told him that this is said to be one of the dark chapters in the Meiji Restoration." When asked about the prime minister's reaction to this, Hatoyama answered, "The prime minister said 'he seems like a great man.' "[59]

Several subsequent writings about Oguri refer to the drama, such as *Learning from the Restoration Years in the Present* (*Bakumatsu ishin ni manabu genzai*), serialized in *Sankei shinbun* from 2009 to 2010. Yamauchi Masayuki begins his paean to Oguri with the claim that people now know Oguri because he appeared on NHK. He acknowledges Oguri's fame through buried-treasure stories, but like the message delivered by memory activists, Yamauchi will hear none of it: "If he is to be famous for money . . . it should be for establishing the foundation for the dollar to *koban* exchange rate at the Philadelphia mint."[60] In *Sankei shinbun*, a conservative pro-LDP newspaper, Yamauchi sought to make the Restoration years a teaching moment for the LDP. The Democratic Party of Japan's (DPJ) victory over the LDP in 2009 was analogous, for Yamauchi, of the Meiji transition, with LDP politicians acting the part of the noble losers. Thus, those who suffered vilification or marginalization, such as Ii Naosuke, Matsudaira Katamori, and Oguri, should be remembered by both the LDP and the DPJ for what they contributed through their sacrifice, and those who succeeded in having a meaningful career after Restoration, men like Fukuchi Gen'ichirō and Narushima Ryūhoku, should be models for defeated LDP politicians.

Oguri's television exposure helped create a market for Kimura Naomi's manga series. Kimura had wanted to create an Oguri manga after discovering Sakamoto Fujiyoshi's biography of Oguri in a used bookstore in Tokyo. He was struck by how different Sakamoto's version of the Meiji Restoration was from what he knew of the event, but at the time, Kimura could not find a manga magazine editor interested in doing an Oguri project. Shortly after Kimura's discovery, Oguri became widely known through treasure hunts depicted on television and praise from Shiba Ryōtarō. It was at this point, sometime during the 1990s, that Kimura and his editor traveled to the Tōzenji to learn more about Oguri.[61] Not until a decade later, however, did Kimura find an interested editor. The senior editor at the manga magazine *Comic Ran Twins* happened to be an Oguri fan, and the monthly series started in January 2005.[62]

In his manga, *A Samurai to the Ends of the World* (*Tengai no bushi*), Kimura contrasts a nationalist version of Oguri against provincial rivals such as Katsu Kaishū and Saigō. Kimura envisions Oguri as an economic nationalist—one who advocates the harmonious unity of all Japanese, regardless of status, against a threatening West. The manga glosses over

Hyōgo Shōsha's failure to materialize; instead, Oguri is portrayed as advocating a status-less economic country unified against foreign pressures. He tells Katsu, "Foreign countries might not try to invade us militarily, but trying to take us over economically is the same."[63] Oguri further argues that Japan's defense is not a military one, and therefore the country cannot rely on samurai to develop Japan; the merchants and others with financial knowledge must lead, working together as one. Oguri confides to a merchant that even he suspects that the Tokugawa shogunate may fall, but says, "The big picture in the world today is the economy. This country will be supported by men like you."[64] He then explains the English word "company" to the merchant in egalitarian terms: "It's an institution that does business comprised of capital from samurai, townspeople, and peasants, with no status distinction."[65]

In Kimura's manga, Saigō appears as a buffoon. Unlike Oguri, Saigō is unable to see the big picture, an economically powerful and unified Japan. Kimura uses a fictional character, the only one in the series, as a device to flip the reader's expectations about the putative losers and victors. A young Satsuma samurai named Miyazato Danjūrō encounters Oguri through a mutual acquaintance and learns about Oguri's vision for a modern Japan where commerce dominates without status. Miyazato reports Oguri's activities to Saigō but undergoes a conversion in the course of the story, eventually abandoning Saigō and supporting Oguri's view of Japan. He faces an identity crisis; on one hand, he has mastered swordsmanship and married into a rustic samurai family, which has allowed him to move up in social status, yet he cannot see how he will fit into the world as Oguri envisions it. He confronts Saigō: "Tell me, as a samurai, what am I to do in the future?" Saigō responds, "You don't have to think; you are a Satsuma samurai, and you must follow heaven's will."[66] Saigō's problem, as Kimura sees it, is that he thinks of himself as only a samurai. When Miyazato questions Saigō's violence against the rival Chōshū domain, Saigō can only react as a samurai: "Today I've come to understand, warriors are warriors exactly because they fight."[67] Miyazato, however, recalls Oguri's words about samurai violence: "Such fighting is not about Japan; it's just a quarrel among samurai."[68] At the end of the manga, which takes place after the Meiji Restoration and Oguri's execution, Miyazato works as a *rikisha* driver. He takes Saigō as a passenger and dumps him onto the ground, ordering him to apologize for killing Oguri. "Much of the Meiji civilization and enlightenment were the very same things Oguri talked about," yells Miyazato, to which Saigō asks, "What kind of world did Oguri want to create?" Miyazato answers, "He only wanted to cooperate and create a country together with men like you."[69] Magnanimous, farsighted, and conciliatory, Kimura's Oguri is a model of patriotism in a Japan experiencing disorder.

Kimura's manga is unique because it includes legitimizing voices that support his interpretation of Oguri and the Meiji Restoration. The first two volumes feature commentary by politicians of the then ruling LDP. These commentaries were not originally included in the serialized version but were added at the suggestion of the book's designer, who was friends with Iwaki Mitsuhide, the deputy chief cabinet secretary. Iwaki agreed to write a commentary, which appears in the second volume, and also contacted former prime minister Mori Yoshirō, who wrote for the first volume.[70] Kimura himself contacted contributors for volumes 3 and 4, both of them acquaintances: Shimizu Masashi, a literature scholar at Japan University, and Tanikawa Akihide, vice president of Tsukuba University and a frequent manga essayist.

Mori's and Iwaki's comments support Kimura's vision of Oguri's history as it relates to politics. Mori points out the necessity of studying the losers to understand Meiji Restoration history, in particular, how contemporary Japan can learn from those samurai who created Japanese identity.[71] He accepts the depiction of Oguri as noble and selfless, noting Oguri's fearlessness when serving the country—a message Mori directs toward politicians and modern-day samurai: "This story begins with Oguri on the ship in a storm, heading towards the United States. Oguri tells the others, 'If we sink right now, well, living and dying, that is the will of heaven.' I think the samurai resolve has something to teach today's politicians."[72] Here, Mori believes that Japan is the boat in the storm, and politicians should strive to improve Japan without concern for political consequences. Mori wrote in the political context of 2005, when LDP rebels challenged the status quo but rarely threatened to leave the party, a fact to which he perhaps alludes when he notes: "What makes this story so interesting is people like Katsu Kaishū, wild men who act as supporting cast, but still work in the progressive faction."[73]

Cabinet member Iwaki Mitsuhide also sees Oguri as a model for Japan's politicians. Oguri was "a Japanese before he was a retainer of the Tokugawa," one who thought of the country first and political loyalties second, and worked for the country every day.[74] Iwaki invokes a classic trope of Japanese identity, that of "honest poverty" (*seihin*)—an idea, says Oguri to a Frenchman in volume 2, "that is one of our beautiful virtues."[75] The concept long existed in Japan to refer to the honest, pure, and impoverished life of the intellectual and artist. In 1992, *seihin* was resurrected in Nakano Koji's best-selling book *The Philosophy of Honest*

Poverty (*Seihin no shisō*), part of the Edo memory of the 1990s. But *seihin* is not just a model to keep the masses content with being poor; for Iwaki, "it has something to teach us politicians in this time of reform."[76] Here Iwaki reminds fellow politicians that greed was the source of political corruption during the 1990s.

Much of Shimizu Masashi's commentary focuses on the story line, but he too emphasizes the political messages in the story. Shimizu acknowledges the story's portrayal of Oguri's foresight, administrative skills, and efforts to unify Japanese merchants against the West, "but this is not just a drama about the creation of a new country. It is a superb human drama [through its illustration of] Katsu Kaishū and Saigō Takamori's hatred and envy of Oguri."[77] Shimizu believes that men like Katsu were more dangerous than the shogunate's enemies. "In every era there are countless people with no talent who hold back those with ability. . . . [I]t was these faceless people that Oguri had to fight before creating a new Japan."[78] These kinds of people block progress by collaborating with the enemy. Even this had a political meaning for readers—the LDP was itself split into factions as it tried to find a way out of the Heisei recession.

Tanikawa Akihide casts Oguri as a man of integrity during a year when integrity was lacking. "The key word that sums up 2007 is 'fakery.' From the world of politics to food companies, Japan has become a country where everything is in doubt. When did Japan become a 'country of fakery'?"[79] This is not Tanikawa's invention; the word "fake" (*itsuwari*) was voted the Chinese character of 2007 by the Japan Kanji Aptitude Testing Foundation.[80] Regional food companies and national confectionary chains, such as Mr. Donut, were discovered to have been selling products with falsified expiration dates or made of recycled materials. One meat-processing company labeled a product "100 percent beef" that was actually ground pork. These scandals were a blow to a nation that had long prided itself on customer service and willingly paid high prices for products assumed to be just a cut above the rest—especially better quality than those ever-suspicious foreign products. Integrity also meant being honest about the past. Oguri, Tanikawa argues, was a man of integrity, standing as far from "fakery" as possible, "and once we look past the 'might makes right' version of history, we can understand that time period [Meiji Restoration years] from another perspective."[81] He ties fakery to an issue long argued by Oguri supporters: that Oguri's history has been blocked by a narrative of history dominated by Satchō. He states, "When we look past the 'victor's history' we can see a different side of those who lived then."[82]

The question remains, How, if at all, have these media treatments affected Oguri's popularity? Kimura claims that 20,000–30,000 units of the first two volumes of *A Samurai to the Ends of the World* were printed, a large amount for any manga. Unfortunately, they did not sell well, and the last two volumes had print runs of about 7,000–8,000, not enough to be displayed in major bookstores.[83] This does not mean that the work failed to popularize Oguri. First, the very publication of the manga suggests that Oguri had "arrived" on the national scene. Politicians, sports and entertainment celebrities, even Nissan CEO Carlos Ghosn have their own biographical manga. Second, Kimura joined the discourse concerning Oguri and the Meiji Restoration, connecting him to an already established network of Oguri fans who could use the manga to promote Oguri to friends, family, and neighbors. This might seem like too small of a scale to affect national perceptions of Oguri, but it was these same small, local networks interacting with well-known writers that initially pushed Oguri into the mass media.

While national media exposure did much to boost Oguri's salience in the national consciousness, memory activists still strove to expand the memory landscape, primarily by convincing residents elsewhere that Oguri was also part of their history. The Yokosuka-Kurabuchi sister-city relationship is the oldest example of this sort of effort. Such networks can be difficult to maintain, especially as smaller towns and villages form conglomerates with larger cities, as has continued to happen throughout the Heisei era. Kurabuchi Village, for example, no longer exists. It and several neighboring towns merged with Takasaki City in 2006, and the sister-city relationship with Yokosuka City was abandoned by the Takasaki mayor.[84] Municipal money is no longer used to send officials to Yokosuka for the annual Oguri-Verny festival. Moreover, since the NHK dramas aired, a sort of "mission accomplished" attitude seems to have spread through Gunma Prefecture officialdom, and enthusiasm for promoting Oguri has waned.

Worries expressed to me by memory activists in 2005 that Takasaki would not support Oguri commemoration, however, have proved unfounded. The Decentralization Law of 2000, which forced localities to become more fiscally responsible and encouraged smaller towns to merge, put municipalities at the center of inspiration for decentralized reform in Japan.[85] In an effort to bridge the identity gap between Kurabuchi and Takasaki and perhaps to benefit from history tourism, Takasaki actively supports Oguri commemoration: Oguri appears on the large illustrated Takasaki map outside of Takasaki Station, Murakami Taiken and others lead tours of Oguri-related sites in Takasaki proper, and more lectures are being held in Takasaki community centers and schools than ever before. More importantly, the mayor committed major funding for Oguri Town (*Oguri no Sato*), a rest stop along Route 406, the major road heading north from Takasaki through Kurabuchi toward the resort town of Kusatsu. Intended as a tourist spot and community center, the building will feature space for Oguri-related objects and Kurabuchi history, a shop for local specialties and agriculture produce, and a rest area.[86]

Successfully expanding Oguri commemoration outside of Gunma depends on the efforts of private citizens—memory activists who contribute to the memory landscape of Restoration losers through which Oguri's legacy can thrive. Aizu-Wakamatsu history groups have hosted Oguri-related lectures, and in 2005, the city featured a temporary display of his possessions. In that same year, Aizu-Wakamatsu and Yokosuka became sister cities due to their mutual connection to Oguri.[87] Oguri's story has also become part of studies conducted by other local historians in places such as Niigata, Fukushima, and Nagano. For example, a Nagano historian wove Oguri into a 2004 biography of Takai Kōzan, a local nineteenth-century entrepreneur, artist, and patron of the renowned Hokusai.[88] Takai once met Oguri to discuss plans for starting a company to conduct trade with foreigners. The advertising blurbs on the front cover of the book state, "Oguri and Takai's dreams for modernizing Japan come into view—Hokusai and Kōzan's prayers are answered." Although Oguri is certainly not as famous as Hokusai, the advertising strategy makes clear that Oguri's name carried enough cultural cachet to help sell books about Takai, a relatively unknown local figure.

Oguri historical memory, reconciliation, and fandom culminated in 2008 at Meiji University, close to his former Edo residence, now a YWCA. Surugadai was a neighborhood occupied by many bannermen like Oguri, but only his residence is marked with a plaque, erected in 1992 after prodding by Murakami Taiken. The idea to reconcile Oguri and Tokugawa began in 2006, when Kurino Yoshio, an Oguri fan and coffee-shop owner in Surugadai, met Tokugawa Tsunenari at a local shrine. The two started talking about Oguri, and a month later, when Kurino visited the Tōzenji, he exclaimed to Murakami, "The fifteenth generation [Tokugawa] Yoshinobu fired Oguri, but the eighteenth generation has acknowledged him." The Oguri Promotion Society and a community organization in Surugadai decided to hold an Oguri exhibition near his birthplace. Tsunenari agreed to give a lecture, but on the opening day, Surugadai officials also convinced him to join a parade of bannerman descendants, Oguri's included, which lasted two hours. As one reporter observed, "He (Tsunenari) seemed confused, but somehow he joined in; Tsunenari just could not refuse the excitement of the crowd."[89] During his lecture, Tsunenari used familiar tropes in praising Oguri for contributing to Japan, but added a new interpretation regarding why Oguri's death was a tragic one. Oguri was killed despite the fact that the Tokugawa shogunate *did not* collapse; political authority simply passed to a new group of people—a smooth transition, Tsunenari claims, that is unique in world history. For the Tokugawa house, it seems, there were no losers after all.[90]

Conclusion

Commemoration of Oguri and the Boshin War has been a legitimizing process. Participation by celebrities such as Dōmon Fuyuji and Hoshi Ryōichi lent an air of authenticity to local events, such as the Kakuno-date conference on the Boshin War, which might have otherwise been seen as an oddly timed and oddly located conference. These same celebrities brought attention to Oguri events in Gunma, and their presence at Oguri Festivals marked the Tōzenji as a required pilgrimage destination for those who want to speak about, act out, or illustrate through manga Oguri's story. Historical memory also created new celebrities: Murakami Taiken is referred to as "Murakami sensei" through his Oguri-related publications and lectures, and his guardianship of Oguri objects. Even Mizuno is a "sensei" to those searching for buried treasure; the documentary about him features a stream of visitors either wanting to offer their theories or seeking guidance on their own treasure quests. But the relationship between national celebrities and native informants is not unidirectional. Local memory activists provide legitimacy and knowledge to Dōmon and Hoshi for their writings about Oguri, and the media seeks these men out when featuring him in documentaries. In exchange, both men interpret Oguri's story in ways that are sympathetic to local narratives.

The difference between the commemorative activities of the Boshin losers and those of the Oguri memory activists concerns their respective visions for Japan. On the one hand, the Aizu descendants' reluctance to reconcile with their former enemies is not just about stubborn local identity. They challenge the conservative assumptions advocated by the postwar government and supported by occupation authorities that later accelerated during the 1980s, a narrative of Japanese uniqueness and harmony that has only recently faltered. In fact, these cross-regional reconciliations celebrate their mutual irritation with the central government. They follow the recent trend found in many late twentieth-century depictions of the Restoration years: looking to the maligned Restoration losers to recover missed opportunities from the Tokugawa period that were rejected in favor of the centralization and imperialism of the early twentieth century.

Not since the 1950s, when Ninagawa used Oguri and Restoration history to attack ahistorical myths about the emperor, has any Oguri commentator fundamentally challenged familiar nationalist tropes. During the 1990s, following the publication of *The Japan That Can Say "No" ("No" to ieru Nihon)*, Murakami Taiken argued that Oguri was in fact the first Japanese to say no to America (at the Philadelphia mint)—a theme appropriated by other authors.[91] And although Oguri is always praised for introducing Western infrastructure and ideas to Japan, Kimura's manga are generally antagonistic toward foreigners: Russians are shown raping local women in Tsushima, the British diplomat Alcock appears evil, and the French are generally depicted as bumbling, except the Yokosuka engineer Verny.

In addition to the mass media, Oguri's rehabilitation is found in small but significant sites of memory. In 2003, he appeared in mainstream education materials for the first time: in a single high school history textbook published in 2003 by an Osaka-based company. Oguri does not appear in the Meiji Restoration portion of the book; rather, he is part of a three-page feature on Yokosuka called "History and Life." According to the instructor's manual, Yokosuka was chosen because it clearly illustrates how history, economics, and international politics have developed over time in the countryside.[92] The student volume contains a picture of Oguri's and Verny's busts and explains that Oguri wanted to build Yokosuka because he believed Japan—not just the shogunate—needed a naval installation.[93] Oguri's biography is contained in the instructor's manual and ends with the following: "In recent years Oguri has been praised for his logical management style, politics, the long-range outlook embodied in his policies, and his way of life. He has been featured in television dramas and the like."[94] Oguri's first appearance in mainstream education may be short, but for memory activists, it represents a significant step since Gonda villagers first tried to rehabilitate him over a century ago.

Conclusion
Meaningful Landscapes

It was the early spring of 2005 but already hot inside the Kurabuchi Village gymnasium in the mountains of Gunma Prefecture. A group of men from the distant coastal city Yokosuka, dressed in traditional *hakama* and formal broad-shouldered vests reminiscent of samurai clothing, began to chant in the slow methodic voice of the Noh theater. They recounted Oguri's story, celebrating him as a reformer and mourning his execution by imperial troops. Only a few years earlier, a curious but no less sincere group of men stood solemnly in the blowing snow as a Buddhist monk chanted in front of a makeshift altar. They hoped that the monk's blessing would help them find the Tokugawa treasure supposedly buried by Oguri on Mount Akagi. After years of planning and weeks of digging, they came to a painful conclusion—there was no gold under the Akagi Club golf course after all.

The juxtaposed scenes might seem unconnected and trivial, but they illustrate the most recent manifestations of a century-long effort to commemorate and invoke the history of one Meiji Restoration loser among many. The two scenes were unlikely to have been viewed by the same audience. The first, a public annual event, seemed almost evangelical—it celebrated local history and identity and attracted tourists. The men from Yokosuka were participants in the memorial landscape that joined Kurabuchi and Yokosuka long ago, from the Yokosuka fiftieth anniversary held in 1909 and the years of cooperation to erect a statue in 1922, to the postwar business and tourist connections. The second scene was a private one, made visible by a documentary filmmaker who spent several years trying to figure out why Mizuno and others could believe in such an unlikely legend. The director never makes Mizuno look like a fool; rather, he imparts the sense that just enough confusion lies at the legend's origin that anyone could convince themselves it was true. Mizuno remained passionate about his cause and earned respect from like-minded adventurers because he sought to recover the reputation of his lost samurai ancestry.

Several poignant silences have marked Oguri's commemoration over time. First, memory activists have rarely, if ever, acknowledged his political weaknesses. Instead, they have shifted the blame for Oguri's failures onto shogunate elites who refused to listen to his proposals. Second, they are largely silent on the inherent tension that Oguri brought with him to Gonda and the surrounding area. Whatever his intentions there, villagers died and houses were burned during the 1868 riots. A proposed monument to commemorate those dead never materialized. Moreover, Oguri's "supporters" did not always selflessly obey their lord. Just as Minomura Rizaemon saved his employer, the Mitsui merchant group, by betraying Oguri, so too did the Gonda Village headman, Satō Tōshichi, act in his own interests, hedging his bets against Oguri by borrowing money from him and eventually becoming the target of other Oguri retainers. Rival memory activists have also become targets for selective memory. A recent anthology called *Everything about Oguri Tadamasa* (*Oguri Tadamasa no subete*), edited by the Tōzenji Temple–based Murakami Taiken, does not mention the Fumon'in Temple in the glossary of Oguri-related historical sites, despite listing other locales with less salient connections to Oguri.

The mass media solidifies a figure's popularity as a national hero, but its reach depends on the work of locals. In one form or another, Japanese people know Sakamoto Ryōma through Shiba Ryōtarō's novels and their manifestations in film and television dramas, including yet another Sakamoto-featured Taiga drama in 2010 (*Ryōma den*). But even Shiba's knowledge about Sakamoto and the Shinsengumi comes from research conducted by a local historian.[1] Local research alone, however, cannot bring a hero to national attention; there also has to be materiality, objects, and physical commemoration. In Sakamoto's case, although local people's rights activist Sakazaki Shiran wrote a fictional account of him in 1883, and in 1904 the empress said she saw Sakamoto in a dream (a story propagated by a Tosa man working in the government), it was not until 1928, after an object was erected locally, that his popularity peaked during the prewar era.[2]

Oguri's commemoration, like that of Ii Naosuke, the Aizu samurai, and others, has been a long and possibly incomplete journey. Ii's early legacy was infamous, and efforts to raise statues to him became sites of conflict. More important than representation and contention, however, is the discourse produced from the event that both supported and attacked him. The textually mediated memory about Oguri established in the early twentieth century continues to represent the core themes in his promotion, but the appropriations have changed over time. His image required more groundwork than did Ii's. Early memory activists needed to educate local citizens about Oguri and create a foundation for promoting him while trying to rehabilitate him at the national level. Former Tokugawa men used his story to attack the actions of both the Meiji oligarchs and other colleagues seen as traitors. Here, memory is a useful lens for examining the often-ignored personal connections among historical personalities. For example, Ōkuma's connection to Yano Fumio, his brother Sadao (who married Oguri Kuniko), and Minomura Rizaemon, or his rivalry against other oligarchs, was not just about politics; personal historical connections also mattered. The same can be said about Fukuzawa's rivalry with Katsu and Tokutomi, or his shared appreciation, with Kurimoto, of the meaning of Oguri's life as a model for the present.

But commemoration is neither straightforward nor teleological. Oguri's popularity fluctuated locally and nationally. In late nineteenth-century Gunma, only the slightest evidence suggests that anyone other than a small group of educated elites cared about his story. Still, this group was dedicated enough to take advantage of the Yokosuka fiftieth anniversary to promote him. With help from allies in the government and in Yokosuka, they succeeded in erecting his statue in 1922, increasing the rate and intensity of local discourse about him. The Restoration boom experienced during the late 1920s through the 1930s was a convenient catalyst for launching Oguri as a national hero, when he was depicted as epitomizing patriotism and having military foresight. The competition there disavows any attempt to view the countryside as a site of harmonious cooperation, but even when the Tōzenji and Fumon'in groups attacked each other, they shared contempt for the treasure hunters.

Moreover, Oguri's story failed to excite national audiences or readers during the immediate postwar period. At the local level, memory activists faced a momentary challenge in their efforts to make Oguri a regional hero, but they overcame this by simply pushing ahead, joining forces with the newly created Kurabuchi Village, and having Oguri's history recognized as important to Gunma and Yokosuka City. The large shift from the Fumon'in Temple before WWII to the Tōzenji Temple after WWII is a testament to the malleability of commemoration, memory, and historical production. By the 1980s, activists benefited from the spreading Oguri memory network, money from the central government, and the *furusato* boom. These developments converged on a shocked public, traumatized during the "lost decade," when a new set of heroes emerged.

During the early 1990s, however, Oguri remained an obscure figure even in Gunma. In a 1994 survey conducted among students at Gunma University regarding which historical figures they recognized, Oguri ranked seventeen out of thirty. Of the students who knew his name, roughly 21 percent had read about him in books or magazines, 14 percent from "elsewhere," and 3.5 percent each from "class" and "television and friends."[3] Oguri had already appeared in the TBS-sponsored gold digs, but it seems that perhaps the written work about him had more impact. By the late 1990s onward, activists, including counterparts who supported the Aizu dead, used their hero to participate in national conversations about the country's direction and the role the past plays in the present. But like the treasure hunters, who always claim that the treasure is only one more meter away, finding solutions to contemporary societal problems by digging up the past seems out of reach. Still, the goal does not matter as much as the process. Mizuno died in 2010 never having found gold, but searching for it was his *ikigai*, it gave his life meaning, and he published books about his family that testify to their legacy.

More important than the sheer number of people who come to know Oguri is the issue of what kinds of people the story reaches. In this case, efforts by memory activists have been successful. In 2005, during a meeting for the privatization of the postal service, it was argued that Oguri created the foundation for the post service before Maejima Hisoka, the father of the Japanese post service; and in 2009, during an investigative meeting for Citizen Welfare and the Economy, politician and writer Sakaya Taichi suggested that the government should follow the example of Oguri's tax, financial, and monetary reforms.[4] The dissenting voices that question how memory activists portray Oguri, such as those who attacked Ninagawa's interpretation or Kiya's book, have faded.

Much of Oguri's popularity came from fantasies about treasure at a time when wealth was disappearing. Memory activists might react negatively toward the buried-treasure legends, but the legends give Oguri publicity. The future of Tokugawa treasure hunting is bleak. Kawahara is getting older, and neither he nor Mizuno has a successor to continue their adventure. Kawahara announced in 2000 that he had found Oguri's Tokugawa gold, right there in Tsukiyone, where his father had begun digging in the 1930s. When the press arrived, Kawahara's son was unenthusiastic: "I'm his son, but this buried treasure talk is horrible. I don't see him, we don't talk much, just leave your business card and I'll make sure he gets it."[5] Kawahara later showed off a gold coin and a bronze sword, which, unfortunately for him, were determined to be objects most likely bought from an antique store, not recovered from underground. Still, those who contributed money to Kawahara's cause did not lose faith.[6] Mizuno continued to dig even after his surgery for lung cancer in 2008. The 2006 documentary portrays him in a sympathetic light. He had no financial supporters, was not desperate enough to fake results like Kawahara, but strove to complete a task started by his grandfather. He was even reluctant to participate in the TBS-sponsored digs, for which the treasure legend did provide ratings gold.

Oguri's commemoration shows no sign of slowing down. Murakami Taiken maintains an active lecture and study tour schedule, and updates Oguri fans through newsletters, email lists, and postings on the Tōzenji Temple's website.[7] He has even introduced Oguri commemoration to the United States and China. Japanese American groups in Los Angeles and San Francisco (May 2010) invited him to give talks about the 1860 embassy. In his lectures, Murakami promoted Oguri. As part of a series celebrating 150 years of Japanese–U.S. relations in 2010, he was interviewed for a short account of Oguri's history, which included the ubiquitous screw, by a New York newspaper catering to Japanese residents.[8] Murakami sent one of Oguri's screws to the Museum of the City of New York exhibition *Samurai in New York: The First Japanese Delegation, 1860*, with the following caption: "This steel screw was one of many brought back by Oguri Tadamasa, the Special Censor, sometimes referred to as the third ambassador. . . . [F]ive years after his return to Japan, he was instrumental in the establishment of a navy yard at Yokosuka, which is regarded as the beginning of modern industry in Japan—property of the Tōzenji."[9] It may seem like an innocuous caption, but the screw, Tōzenji, and Oguri's status as a progenitor of Japan's modernity are products of commemoration that date to the late nineteenth century. Oguri-as-modernizer also interested a producer from the Chinese Business Network (CBN), who interviewed Murakami in 2012 about Oguri, in particular his financing of Yokosuka. Both Oguri and Murakami later appeared in CBN's show "Golden Age."

Murakami and Oguri fans have never given up on the idea of a feature-length Oguri film, and it remains a possibility. In November 2009, Murakami told newsletter subscribers that director Cellin Gluck suddenly appeared at the Tōzenji with a small group from a Japanese film studio. Gluck, a Japanese American, was in Japan for the release of his Japanese remake of *Sideways* and only the day before had given the film's opening-day greetings in Tokyo. It is unclear who initiated the visit to Tōzenji, or how Gluck discovered Oguri's story, but the movie never made it past the discussion stage.[10] If a major Oguri film is produced, we will be able to see for ourselves just how far memory activists are able to project their interpretations into the narrative, and whether buried treasure will become a promotion hook. Ultimately, the goal of memory activism is to naturalize the association between the object of commemoration and its memory landscape, to the point where the object is no longer seen as the result of commemoration. To appropriate, in my own way, a phrase by Kōjin Karatani, "once a landscape has been established, its origins are repressed from memory."[11] In Oguri's case, memory activists want the public to see Oguri as just part of history and not part of commemorative artifice to make him popular. Memory activists have brought him close to this point, but have not quite reached it. For example, while the student version of the history textbook discussed in chapter 5 naturalizes the connection between Oguri and Yokosuka, and the importance of both, the instructor's version notes that Oguri's popularity is only a recent phenomenon—the role of local boosterism is still visible. If memory activists succeed in making Oguri a truly national figure, they will have also reshaped how the public interprets the Meiji Restoration.

Finally, memory landscapes do more than act as networks for commemoration; they lay the groundwork for other, more meaningful connections. When Oguri's granddaughter and her family needed to escape the Tokyo firebombing, she took refuge in Gunma. So too did military men from Yokosuka. Their only connections to Gunma, more specifically the Gonda area, were commemorative, not personal ones.

The most powerful memory landscape that has recently emerged in this way is the one that has helped funnel assistance to the northeast that still suffers from the 3.11 disasters. Up until 3.11, the cities of Aizu-Wakamatsu and Hagi seemed as though they would never fully reconcile or become sister cities, despite ongoing educational exchanges and mutual tourism. Their interaction suggests that national identity is weak, as if there were no other way to relate to each other, other than through this shared early modern history. By the twenty-first century, no national

goal had brought the two regions together besides their historical antagonism. But the groundwork established through this memory landscape,
first by memory activists and then by city hall officials, facilitated Hagi
City's reaction to the 3.11 disasters in the northeast. In April, Hagi City sent
twenty million yen, plus supplies, to Aizu-Wakamatsu. The Hagi mayor
commented, "We've had this relationship since the Boshin days, but now
we face a national crisis that must be overcome."[12] In the fall, following a
conversation between the two mayors during the summer, fifty-five Hagi
citizens accompanied their mayor on a tour of historical sites connected
to the *byakkotai* in Aizu-Wakamatsu. "Historical facts are facts, but we
need to move forward into the next generation and develop exchange
between us."[13] In both cases, historical memory was invoked to create an
emotional bond between the two cities, with Hagi acting as the recon-
ciler for past wrongs. Of course other prefectures and cities without any
historical connection to the northeast also donated money, but the Hagi
City mayor's poignant gesture of visiting Aizu-Wakamatsu at a time
when tourists avoided the northeast could only be meaningfully delivered through a memory landscape.

The Aizu memory landscape has become linked to national memory
in 2013 through NHK's Taiga Drama, *Yae no Sakura*. Not long after 3.11,
NHK decided to focus its 2013 Taiga drama on Fukushima prefecture
(Aizu) not only to drive tourism to Fukushima, which is estimated to
bring in some 11.3 million dollars' worth of yen, but also to change Fukushima's association with disaster to that of Yae and Kakuma.[14] This assumes, of course, that the association between Fukushima and Yae and
Kakuma will be a good one. The story follows the life of Yae, wife of the
famous educator Niijima Jō. Her father was the Aizu domain's gunnery
instructor and her older brother, Kakuma, eventually inherited the position and became caught up in the Boshin War. Yae also led a small group
of young samurai to fight against the emperor's army.

And here lies one challenge of tying a local memory landscape to a
national one. As one commentator has noted, NHK tries to create programs that include everyone in Japan (*minnasama no NHK*). But how
will the producers of *Yae no Sakura* avoid offending those in western
Japan, whose legacy is tied to the victors of the Boshin War, the same
people who persecute Yae and Kakuma, the stars of the current drama?
So far, viewer ratings have suggested that viewers in distant parts of south-
western Japan simply are not interested in Aizu memory. The viewer ratings for episode seven (February 2013) are as follows: Fukushima, 27
percent; eastern Japan, 17.5 percent; western Japan, 15.9; and in northern
Kyūshū where many of the Meiji Restoration heroes are from, only

10.3 percent.[15] One thought is that perhaps the drama will appeal to south-westerners by highlighting Yae's relationship to Jō, who built Dōshisha University in Kyoto, whose graduates are affectionately known as "*dō-yan*" in western Japan.[16] Either way, it seems that even in places where Yae and Kakuma's legacy is thin, such as Annaka city in Gunma, Yae's husband's natal home, local cities are taking advantage of their spot on the memory landscape, selling Yae goods and Fukushima products, thus connecting themselves to Meiji Restoration memory and doing so for a greater cause.

Notes

Introduction

1. This, of course, was a problematic construct; both sides believed they had been fighting on behalf of the emperor, and many participants were in it for the adventure or honor rather than out of fanatic support for one cause over the other.

2. Fine, *Difficult Reputations*.

3. McClain, *Japan: A Modern History*, appendix 29.

4. Tanaka Satoru recently filled the lacunae in Aizu and loser identity with his *Aizu to iu shinwa*.

5. "Memory activist" is Carol Gluck's term.

6. Gillis, ed., *Commemorations*, 5.

7. Gluck, "The Invention of Edo."

8. Saigō's commemoration started with Ivan Morris, *The Nobility of Failure*, but more recent essays include Berlinguez-Kōno, "How did Saigō Takamori Become a National Hero After his Death?" and Ravina, "The Apocryphal Suicide of Saigō Takamori." Ravina's article builds on similar Saigō memory projects in Japanese, such as Ikai Takaaki, *Saigō Takamori*; Ikai Takaaki, "Shizoku hanran to Saigō densetsu"; and Sasaki, "Saigō Takamori to Saigō densetsu." On Restoration booms, see Gluck "The Past in the Present." Saaler and Schwentker see their edited volume as a first step toward a Japanese version of Pierre Nora's project. See Saaler and Schwentker, *The Power of Memory in Modern Japan*, 6.

9. Narita, *Shiba Ryōtarō no bakumatsu Meiji*.

10. Connerton, *How Societies Remember*, 6.

11. Fine, *Difficult Reputations*, 7.

12. Confino, "Collective Memory and Cultural History," 1393–94, 1403. Confino has in mind the works of Gildea, *The Past in French History*, and Rousso, *The Vichy Syndrome*. While acknowledging the importance of both works, he argues that Gildea's study of collective memory tends to focus too much on the political memory of Catholicism, liberalism, socialism, and so on, constructed by party leaders, intellectuals, statesmen, journalists, and politicians. In Rousso's work, Vichy memory appears too top-down in its focus on the figure of de Gaulle and on public and official memory.

13. Hue-Tam Ho Tai, "Remembered Realms," 913.

14. See, for example, Peterson, "History, Memory and the Legacy of Samori in Southern Mali, c. 1880–1898," and Troyansky, "Memorializing Saint-Quentin." Some scholars also refer to local memory sites as "counter-memory." See Lipsitz, *Time Passages*.

15. Fine, *Difficult Reputations*, 22.

16. Itō, "Ishinshi no kakinaoshi," 2.

17. Rieger, "Memory and Normality," 564. Communicative memory has been cited as one way to move forward in memory studies.

Chapter One

1. Ryōhei Koitabashi, *Katsu Kaishū no raibaru*, 60.

2. *Gunmaken shiryōshū 7*, 8. There are only two portions of his diary dating from 1/1867. Four account books exist for 1850, 1858, 1860, and 1862; they have been republished in *Gunmaken shiryōshū 7*.

3. Schwartz and Schuman, "History, Commemoration, and Belief," 185.

4. Nakajima Mineo has pointed out that many of the retainers who held influential positions in the shogunate either were low-ranking samurai or, like Katsu Kaishū and Fukuzawa Yukichi, were originally from commoner stock. According to Nakajima, Oguri was one of the few talented retainers who happened to be from a family with deep connections to the Tokugawa. See Nakajima, *Bakushin*, 188–89.

5. *Tokugawa jikki 1*, 147–48. The current heir is Oguri Mataichirō Kazumata. He drew the characters Perurin and Ogurin for the 150th anniversary of U.S.-Japanese relations celebrated in Yokosuka.

6. The rest of the Oguri heirs were buried in the Hō'onji in Edo, the family temple of the main branch.

7. Yamamura Kozo's study places the total number at roughly five thousand in the early eighteenth century and six thousand in the latter part of the Tokugawa period. See Yamamura, *A Study of Samurai Income and Entrepreneurship*, 3–4. The typical bannermen only received between 500 to 600 *koku*. According to Kawamura Masaru, only 36 percent ranked above 1,000 *koku* and less than half possessed their own fief lands. See Kawamura, *Hatamoto chigyōsho no shihai kōzō*, 1.

8. Kawamura, 1.

9. *Sanōshi-shi*, 13. The family possessed 2,500 *koku* worth of land in four provinces—Kōzuke, Shimotsuke, Shimosa, and Kazusa—all located outside of Edo.

10. See Nakako Fukui, "Kōkei no chigyōjo no shihai," in *Kinsei kokka*.

11. Ibid., 32.

12. The complete document and details of an attempt to send villagers to Edo, and how they were stopped by local monks, are in *Sanō-shishi*, 496–498.

13. Ibid., 496.

14. Ibid.

15. *Kurabuchi sonshi*, 157. Gonda villagers were not involved in the protest. Unfortunately, there are no records describing how Gonda villagers perceived their own tax rate, but they might have been aware that theirs was one of the highest in

the area. As Neil Waters pointed out in his study of villages in Kawasaki, a tax hike in one village could result in a region-wide crisis because the newly burdened village might pull out of cooperative projects. See Waters, *Japan's Local Pragmatists*, 42.

16. Wigmore, *Law and Justice in Tokugawa Japan,* Part 8-B, 30–33.

17. Koitabashi, 20.

18. Tadataka was the son of Nakagawa Tadahide. Tadakiyo had another son, by natural birth, before dying at age twenty-two. That younger son, later known as Kazuma, was no longer needed, and he married into another bannerman family—the Kusaka.

19. On Satō Tōshichi, see Koitabashi, 43–48. He is also said to have personally delivered the annual tax to the Oguri family, walking the entire way to Edo. See Koitabashi, 45. Satō married the younger daughter of Oguri's other Gonda retainer and member of the village elite, Nakajima Sanzaemon. The Nakajima family eventually moved to Tokyo and took Satō's diary from the 1860 mission with them.

20. Roberts, *Mitsui Empire,* 65. Shibusawa Eiichi also noted Kimura-Minomura's illiteracy: "take someone like Minomura Rizaemon at Mitsui, for example; that man could barely read." Shibusawa, *The Autobiography of Shibusawa Eiichi*, 137.

21. *Mitsui ginkō hachijūnenshi*, 45.

22. Nakamura, *Satō Issai, Asaka Gonsai*, 146.

23. Murakami, *Oguri Tadamasa no subete*, 49.

24. Tokugawa retainers included Kawaji Yoshitarō (foreign commissioner), Kurimoto Joun (who studied at the Oguri residence), Fukuchi Gen'ichirō (Meiji newspaper giant), Nakamura Masanao (later novelist and writer for the *Meiji Six*), and Kimura Kaishū (leader on the *Kanrin Maru* that sailed to America in 1860). Non-Tokugawa retainers: Yoshida Shōin, Takasugi Shinsaku, and Iwasaki Yatarō (founder of Mitsubishi).

25. DNISK, AN 156-0006, 0007.

26. Kikegawa, "Gaikōkan to shite no Oguri Tadamasa," 15. The assassins went undiscovered and cost the shogunate $10,000 in indemnity payments. See DNISK, AN 159-0103 for more on the murder.

27. For example, Miyoshi Masao, *As We Saw Them*.

28. Article five of the Harris Treaties provided that Americans and Japanese might use either Japanese or foreign coin to make payments, and all foreign coins would pass in Japan for Japanese coins of equal weight. In addition, for a period of one year after the signing of the treaty, the Japanese agreed to exchange their coins for foreign ones based on equal weights with no barriers limiting the exportation of Japanese *ichibu* silver coins. The Mexican silver dollar used for trade by Westerners throughout much of the world weighed three times more than the *ichibu* yet was 20 percent less pure in its silver content. Moreover, the ratio of gold to silver was 1:5 in Japan but 1:15 in the West. Merchants freely exchanged one Mexican dollar for three *ichibu*, traded silver for relatively cheap gold, and then sold the gold back in the West at a profit. After several unsuccessful attempts to debase Japanese coins by reducing their bullion content, increasing their weight, or creating new coins, the shogunate finally reached an agreement with the Western countries to mint a new coin for use in foreign trade, but this did not become a reality until 1872. See Peter K. Frost, *The Bakumatsu Currency Crisis*. For more on Oguri's participation in cross-commissioner

miscommunication, see Roy S. Hanashiro's account of the shogunate's attempt to establish a mint in Hanashiro, *Thomas William Kinder and the Japanese Imperial Mint*, 29–32.

29. Oguri had been married by this point in his career, but the exact date remains unknown. According to his 1858 account book, Oguri also employed two concubines.

30. Satō recorded his experience in a diary covering the entire trip, from pre-departure on 1/18/1860 to the return of the embassy on 8/29/1860. It and his notes were recently published with a modern Japanese translation and commentary. See Murakami Taiken and Satō Tōshichi, *Bakumatsu kenbei shisetsu*. His diary contains sketches, but given the detail of these pictures, it is believed that another member of the embassy drew them, not Satō. There are two copies of his diary, one kept at the Tōzenji Temple.

31. DNISK, BU 011-0721. *Owazu sōrōwaba sessha wo teppō niha uchihatasu sōrō to mo zonnen shidai ni itasubeku sōrō.*

32. Ibid., BU 016-0015.

33. Ibid., BU 017-0362.

34. DNISK, BU 017-0935.

35. DNISK, BU 017-0932.

36. Totman, *The Collapse of the Tokugawa Bakufu*.

37. Murata and Sasaki, *Zoku saimu kiji* 1, 91–92.

38. Totman, 38.

39. DNISK, BU 089-0694.

40. Ishii, *Meiji ishin no kokusaiteki kankyō*, 190. A report (*fusetsusho*) from the Higo domain describes the plan: one thousand infantry, eight cannon, and one hundred cavalry to ask the court to allow trade with foreigners to continue. The diary entry by a Mito samurai also describes the plot, adding that "Oguri and others" wanted to crush the emperor. See Ishii, 191. These failed plans might have worked under the then senior councilor (*tairō*), Ogasawara Nagamichi, who sent troops that almost reached Kyoto for this very purpose before being turned back.

41. The only academic study of Yokosuka history in a Western language is de Touchet's *Quand les Francais armaient le Japon*. For a brief history of American fleet activities in Yokosuka, see Tompkins, *Yokosuka: Base of an Empire*.

42. This last point was made to me by personal communication with Oguri Research Society (Oguri Kenkyūkai) president Murakami Taiken.

43. DNISK, BU 011-0663.

44. DNISK, GE 045-0289.

45. Broadbridge, "Shipbuilding and the State," 602.

46. From more on Léon Roches, see Lehmann, "Léon Roches," 273–307. For more on the French role in the shogunate, see Medzini, *French Policy in Japan during the Closing Years of the Tokugawa Regime*, and Sims, *French Policy Towards the Bakufu*.

47. Ishii, 621. Some Oguri biographers deny that Yokosuka was meant to serve any military purpose during the *bakumatsu* period, arguing that Oguri's intentions were purely economic. However, the foundry was clearly intended to strengthen the shogunate's military in the short term. The shogunal elders were also aware of Roches's own military experience. See Ishii, 624.

48. Takahashi, *Yokosuka zōsenjo zōsetsu*, 3.

49. Kurimoto had been exiled to Hakodate in the early 1850s, during which time he met Mermet de Cashon, a French priest stationed in Hakodate. The two men exchanged language lessons, and Cashon later became Roches's interpreter and secretary. Roches, who often acted on his own without complete supervision from the French government, hoped to solidify French interests in Japan by helping the shogunate build the Yokosuka foundry.

50. Ericson, "The Bakufu looks abroad," 386. The school taught a broad range of subjects, including math, science, and even French literature. In addition to shipbuilding, the French at Yokosuka helped the shogunate start other industries, such as the cultivation of vegetable bitumen used in waterproofing. But it was Oguri who had the foresight to follow through on French recommendations, and he acknowledged Verny's help in this regard. See Hashimoto, "Introducing a French Technological System."

51. Tanaka, Kuwabara, eds., *Hirano Yajurō*, 58–59.

52. Roberts, 66.

53. Ibid., 70–71.

54. Ibid., 68.

55. Ibid., 77.

56. Ibid., 78.

57. Ibid., 79.

58. Motoyama, *Proliferating Talent*, 33.

59. Takai Kōzanden Hensan Iinkai, *Takai Kōzan den*, 213–20.

60. Ibid., 228–29.

61. Infrastructure included such items as gas lamps and a post office. In addition, the group was supposed to issue its own paper money to be backed by each individual merchant's capital. See Tokuda Atsushi, "The Origin of the Corporation in Meiji Japan," 4.

62. DNISK, KE 130-0336.

63. Sakamoto, *Oguri Kōzukenosuke no shōgai*, 423.

64. Roberts, 80. Scholars debate whether Hyōgo Shōsha can be properly defined as a company, calling into question Oguri's influence over modern Japanese business history. Tokuda Atsushi believes that it was not a company in the true sense because it lacked free buying and trading of stock, nor was the capital pulled together under the company name, and should therefore be considered a "guild-like union" (*nakama kumiai*) (Tokuda, 5). Sakamoto Fujiyoshi argues that the Hyōgo Shōsha was Japan's first joint stock company because the proposal itself includes the word "company" (*conpenii*), the first such occurrence in any document in Japan. He makes an important statement about language; Oguri used the word "*shōsha*" followed by "*compenii*" to refer to this operation, which differs from the modern use of the word "*shōsha*," which emphasizes the importing and exporting of goods—in other words, a trading group. But because Hyōgo Shōsha also printed and distributed money, combining trade with financial operations, we should think of Hyōgo Shōsha as a *kaisha*, a word meaning "company" and "corporation" (Sakamoto, 423–24). Although the company never began operations, Sakamoto notes that two early Meiji companies in Osaka borrowed much from Oguri's plan, and the merchants assigned to Hyōgo Shōsha went on to direct both of them (Sakamoto, 469).

65. Taken from Walthall, "Edo Riots," 425–26. See also Minami, *Bakumatsu Edo shakai no kenkyū*, 296.

66. A rumor started by the *Daily Advertiser* that both Oguri and the shogun had been murdered was repudiated by a statement in the *London and China Telegraph*: "We have much pleasure in announcing that the Commissioner of the Treasury—Oguri Kodzuk noski . . . is alive and in constant communication with the British Legation" (*London and China Telegraph*, 4 May 1867, 234).

67. Sims, 66–68.

68. Ihara, *Hijikata haku*, 99.

69. Ibid., 398.

70. Noguchi, *Edo wa moeteiruka*, 399.

71. This incident was confirmed by Kimura Kaishū during an interview with Taguchi Ukichi contained in *Kyūbakufu*, a journal that recorded oral histories of former shogunal officials. See "*Kōzuke shidankai dangata*," *Kyūbakufu* 3 no. 3, 49.

72. *Oguri nikki*, 69.

73. Totman, 439.

74. Minomura, *Minomura Rizaemon den*, 52. The relationship between Oguri and Minomura was made into a novel; see Takahashi, *Nippon taihen*.

75. Koitabashi, 13–14. He also taught Lafcadio Hearn, who mentions him in *Out of the East* as one of his teachers of Chinese learning. See Hearn, *Out of the East*.

76. On the Shōgitai and Shibusawa, see Steele, "The Rise and Fall of the Shōgitai."

77. Hayakawa's interviews with Gonda villagers and others who knew Oguri provide us with a rich source of information. Much of his research was published before WWII in the local journal *Jōmō oyobi Jōmōjin*.

78. *Kurabuchi sonshi*, 240.

79. Tamura, *Yonaoshi*, 145.

80. Originally in "Takasakishi shikenkyū" no. 17, quoted in Ichikawa and Murakami, *Oguri Kōzukenosuke*, 77.

81. Shirayanagi, "Oguri Kōzukenosuke ibun," 10–11.

82. *Oguri nikki*, 73.

83. For another example of a bannerman moving to his fief in nearby Shimotsuke Province during this period, see chapter four of Steele, *Alternative Narratives*, 2003.

84. Sugiyama, "Hito hatamoto no Meiji ishin," 7.

85. Ibid., 6–9.

86. Yamamura Kōzō notes that for a little over 93 percent of the bannermen listed in *The Kansei Revised Samurai Genealogies*, incomes fell or remained constant during their lifetimes. See Yamamura, 38. His study traces the deteriorating financial situation of bannermen in the latter part of the Tokugawa period. For another tale of bannerman woe, see Yujiro, "The Reality behind Musui Dokugen," 289–308.

87. Steele, *Alternative Narratives*, 43–49.

88. Nishiwaki, *Hatamoto Mishima Masakiyo*. See also Sugiyama for more on the same bannerman.

89. Morris, "Hatamoto Rule," 12.

90. Kodama, Nishigaki, Yamamoto, and Ushuki, eds., *Gunmaken no rekishi*, 253.

91. Ibid., 254.

92. For more on this phenomenon, see David Howell, "Hard Times in the Kantō," 349–71.

93. Ibid., 261.

94. Tamura, "Jōshū yonaoshi to Oguri Kōzukenosuke," 96.

95. Shibusawa Eiichi and Teruko Craig, *The Autobiography of Shibusawa Eiichi*, 21.

96. Sippel, "Popular Protest in Early Modern Japan," 1977.

97. Steele, *Alternative Narratives*, 45.

98. Suda, *"Akutō" no jūkyūseiki*, 15–17. By analyzing descriptions of uprisings, Suda notes greater use of the word "wretch" (*akutō*), which refers to young peasant men guilty of violent protest.

99. Takasakishi, *Shinpen Takasaki shishi shiryōshū 5, kinsei 1*, 741.

100. Nakajima, *Jōshū no Meiji ishin*, 58.

101. *Oguri nikki*, 78. See also *Eidai kiroku cho* in *Shinpen kurabuchi sonshi 1, shiryōhen 1*, 622.

102. *Kurabuchi sonshi*, 241–42.

103. Koitabashi, 23.

104. Yamaguchi, ed., *Nakanojō machi bakumatsu*, 16.

105. Such sources include but are not limited to *Kurabuchi sonshi* and works by local historians Koitabashi and Ichikawa. The bribe offer is not recorded in Oguri's diary.

106. *Oguri nikki*, 78.

107. Oguri records the number at two thousand, but Koitabashi notes that the Gunma Prefecture and Gunma County histories place the number at seven hundred (Koitabashi, "Oguri no maizōkin," 28). The *Eidai kiroku cho* puts the number at one thousand. See Takasakishi, *Shinpen Kurabuchi sonshi, dai1kan shiryōhen 1*, 622.

108. *Oguri nikki*, 78–79.

109. Short biographies of these men are in Koitabashi, "Oguri no maizōkin," 247–57.

110. Ibid., 79.

111. The Kurabuchi Village history states that eleven houses and two temples were burned. See *Kurabuchi sonshi*, 244. The record of the Iwakōri Village headman states thirteen houses were burned. Gunma kenshi Hensan Iinkai. *Gunma kenshi. 20, shiryōhen. 10, kinsei 2, seimo chiiki. 2*, 809.

112. *Kurabuchi sonshi*, 247.

113. "Eidai kiroku cho" in Takasakishi, *Shinpen Kurabuchi sonshi, dai 1 kan shiryōhen 1*, 622.

114. *Oguri nikki*, 79.

115. Koitabashi, *Katsu Kaishū no raibaru Oguri Kōzukenosuke ichizoku no higeki*, 39. This interpretation is based on the last entry for 3/4/1868: "As a matter of course I had one official from each village stay the night" (*mottomo muramura yakunindomo ha hitorizutsu, tomarioku sōrō*).

116. On "bodies-as-signs," see Botsman, *Punishment and Power*, 19.

117. *Gunma shiryōshū 5*, 15.

118. Yamaguchi, ed., *Nakanojōmachi bakumatsu no uchikowashi to Oguri Kōzukenosuke*, 31–32.

119. Ibid., 33.
120. *Oguri nikki*, 80.
121. Ibid., 82.
122. Ibid., 80.
123. Ibid., 82.
124. *Takasaki-shi shi*, 745.
125. Ibid., 748.
126. Ibid., 749.
127. Ibid., 752.

128. *Oguri nikki*, 83. There is another theory that Oguri's men packed gunpowder into a barrel, set it on a cart and ignited it in order to scare off some of the less committed rioters.

129. On this same day the village headman from Shimosaida visited Tōzenji to ask about Oguri's situation. Because Shimosaida was now under the jurisdiction of Takasaki, the headman probably had advance knowledge of the troop movement and may have informed Oguri. *Oguri nikki*, 85.

130. Ibid., 86–87.
131. Ibid., 87.
132. *Takasaki shi shi*, 735.
133. *Kurabuchi sonshi*, 256–57.
134. Koitabashi, *Katsu Kaishū no raibaru Oguri Kōzukenosuke ichizoku no higeki*, 96.
135. Ibid., 95.

136. Interestingly, in an article published in *Kyūbakufu*, the former *rōjū* Inoue Masanao stated that Oguri was killed by "natives" because he was hated by Kantō sake brewers for taking taxes from them (Inoue, "Inoue Masanaokun kyūjidan," 45). Another rumor suggested that a Hikone samurai beheaded Oguri for advocating a partial reduction of the Hikone domain lands. See Togawa, "Oguri Kōzukenosuke," 36.

137. Shirayanagi, "Oguri Kōzukenosuke ibun," 20. Tomosada, only nineteen years of age at the time, was accompanied by his thirteen-year-old brother. Both were sons of the famous Meiji statesman Iwakura Tomomi.

138. Abe, *Kaigun no senkusha*, 174.
139. Ibid., 168.

140. *Gunmaken shiryōshū 7*, 298. Here "children" could be read as referring to Oguri's adopted daughter (Mataichi's wife) and his unborn child.

141. She was also supposed to act as a body double for Oguri's wife. See Koitabashi, *Katsu Kaishū no raibaru Oguri Kōzukenosuke ichizoku no higeki*, 90.

142. Yoshiimachi, *Yoshii chōshi*, 688.

143. Koitabashi, *Katsu Kaishū no raibaru Oguri Kōzukenosuke ichizoku no higeki*, 159.

144. Ibid., 102.

145. Her concern was legitimate, because a man from Shimoda's village had been killed by Oguri's soldiers during the riot in Gonda. Although Maki might not have known this, anti-Oguri feelings in the area were palpable (see Koitabashi, *Katsu Kaishū no raibaru Oguri Kōzukenosuke ichizoku no higeki*, 107–10). The source for

details of this incident was the village headman's youngest child and Shimoda's grandchild.

146. Ibid., 114.

147. Ibid., 185. Some records put her birth on 6/14/1868, but according to her family register (*koseki*), she was born on 6/10/1868. See Koitabashi, *Katsu Kaishū no raibaru Oguri Kōzukenosuke ichizoku no higeki*, 186.

148. *Shizuoka kenshi shiryōhen 16 kingendai* 1, 54.

149. *Kigo ni noshi wo somedasumo*, literally "even if someone colors the wrapping with their own mark." See Kurimoto, *Hōan ikō*, 104.

150. Ninagawa, *Ishin zengo no seisō to Oguri Kōzuke no shi*, 144. Tatebe was a 15,000-*koku* daimyo of the Hayashida Domain, located in present-day Hyōgo Prefecture.

151. Togawa, "Oguri Kōzukenosuke."

Chapter Two

1. Certeau, *The Writing of History*, 273.

2. Ibid., 272.

3. Shimane, *Tenkō*, 23. Indeed, the only nationally circulated monograph written about Oguri was a young-adult reader published in 1901.

4. Wertsch, *Voices of Collective Remembering*, 5.

5. Jansen, "Resurrection and Appropriation," 962.

6. Assmann and Czaplicka, "Collective Memory and Cultural Identity," 125–33.

7. Translation in Mehl, *History and the State*, 1. For more on the development of professional history writing in Japan and Meiji historiography, see also Brownlee, *Japanese Historians*.

8. Mehl, 39–40.

9. Ibid., 13.

10. Ibid., 161.

11. Numata, "Shigeno Yasutsugu," 264.

12. Ibid., 282. Within the Meiji leadership, Iwakura Tomomi, feeling that the *Dai Nihon hennenshi* was taking too long to complete, and dissatisfied with its simple chronological approach, went so far as to order a shorter history be written to focus on the imperial line, with the goal of securing the emperor's supremacy over an ever more active parliament. However, historians in the Office of Compilation, which Iwakura created, eventually abandoned the project after Iwakura's death. See also Mehl, 29.

13. Calman, *The Nature and Origins of Japanese Imperialism*, 144–45.

14. Ōkubo, *Nihon kindai shigaku no seiritsu*, 278.

15. Mehl, 61.

16. Ibid., 12.

17. Crane, "Writing the Individual," 1375.

18. Nagai, "Meijiki ni okeru kyūbakushin," 47.

19. See Totman's essay on the Meiji bias in Totman, *The Collapse of the Tokugawa Bakufu*, 550–64. Totman states that Oguri, along with Matsudaira Katamori and

Ogasawara Nagamichi, were the most vilified figures badly in need of rehabilitation (ibid., 560).

20. Totman, 561.

21. It has been suggested that Meiji period histories with nationalist overtones, both pro-shogunate and anti-shogunate, emphasized the effort to unify the Japanese people. In so doing, they downplayed the help received from Westerners on both sides and minimized any foreign incidents that occurred after Perry's arrival, such as the Tsushima Incident. This might explain why Oguri, whose career was defined by his relationship with Westerners, was often minimized or neglected. See Rekishigaku, ed., *Meiji ishinshi kenkyū kōza*, 1:11.

22. Huffman, *Creating a Public*, 41.

23. For examples, see *Naigai shinpō* 30, no. 4, 5/4/1868, in Kimura and Meiji Bunka Kenkyūkai, *Bakumatsu Meiji shinbun zenshū*, 4:163; and *Naigai shinpō*, 14/3/1868, in ibid., 4:123.

24. Ibid., 3:335.

25. Ibid., 4:131.

26. Oguri never mentioned Niemon in his diary, although other Oguri relatives—including Nizaemon, Hanemon, Amitarō, Nagayoshi, and Kōtarō—are each mentioned once, and Oguri Shimōzuke is mentioned four times.

27. A *Nichi Nichi shinbun* article stated that in addition to his household goods, the army confiscated 10,000 *ryō* (Kimura and Meiji Bunka Kenkyūkai, *Bakumatsu Meiji shinbun zenshū*, 3:335).

28. This translation is from Fukuchi's version in *Kōko shinbun* (ibid., 3:43). The opening phrase in *Nagai shinpō* is "*watashi honke Oguri Kōzukenosuke*," which could mean "Oguri Kōzukenosuke of the main branch of my family" (ibid., 3:148). The same heading is printed in the *Soyofuku kaze* version (ibid., 3:388). The *Chūgai shinbun* version starts "*moto onkanjō bugyō aitsutome sōrō Oguri Kōzukenosuke*," or "Oguri Kōzukenosuke, who worked as the former honorable minister of finance" (ibid., 3:286).

29. *Nagai shinpō* printed part of the Tōsandō army's arrest warrant, which includes reasons for Oguri's arrest, namely that he was plotting against the government (ibid., 4:180–81). *Nichi Nichi shinbun* states that there were rumors about Oguri building a fort in Gonda, suggesting that this was a reason for his arrest (ibid., 3:335).

30. Steele, "Edo in 1868," 136. Oguri had informed his subordinate Katsu Kaishū months earlier in 6/1866 of a similar but secret plan to eliminate the threat from Chōshū and Satsuma using French military aid, then dissolve the domains and create a centralized political system (*gunken seidō*) led by Yoshinobu. See Steele, "Katsu Kaishū," 146.

31. Shimane, *Tenkō*, 164–65. This may also suggest that Oguri abandoned his earlier hope to eliminate Satchō and create a centralized system as he outlined to Katsu Kaishū. If this is true, then it would also support the theory that Oguri did not intend to resist the imperial army in Gunma, as has been suggested by some scholars.

32. For an outline of Fukuchi's three alternative military strategies, see Huffman, *Politics of the Meiji Press*, 212–13.

33. Ibid., 50.

34. Kimura and Meiji Bunka Kenkyūkai, *Kōko shinbun* in *Bakumatsu ishin shinbun zenshū*, 4:43-44.

35. Ibid., 4:47. Oguri's executioners and other witnesses claim he made no statements before his execution except a request that his family be left unharmed. Many years after the newspaper folded, Fukuchi admitted that he wrote most of the "unknown author" articles published in *Kōko shinbun*, and we can assume this was true for both the commentary above and another Oguri poem. On Fukuchi's anonymous writing, see Huffman, *Politics of the Meiji Press*, 214, footnote 34.

36. Botsman, *Punishment and Power*, 71.

37. Nearly 31 percent of the Meiji bureaucrats in the years immediately following the Restoration had previously served in the Tokugawa shogunate. The breakdown of ex-Tokugawa men by ministry is as follows: 44 percent of the Finance Ministry (ōkurashō), 43 percent of the Grand Council of State (dajōkan), 38 percent of the Foreign Ministry (gaimushō), 37 percent of the Imperial Household Ministry (kunaishō), 34 percent of the Navy Ministry (kaigunshō), 32 percent of the Education Ministry (monbushō), and 13 percent of the Army Ministry (rikugunshō) (see Mino, "Kindai ikōki").

38. Kimura and Meiji Bunka Kenkyūkai, *Kōshi shinpō*, 5/21/1868, 111.

39. Huffman, *Creating a Public*, 44.

40. Thousands of Tokugawa retainers disappeared from the historical record, but most of those who attained some degree of national fame were employed in posts that involved them with foreign affairs. For example, most of those active in the Meiji Six Society (*Meirokusha*) had worked in the Western Languages Study Institute (*Kaiseijo*) (Mertz, *Novel Japan*, 90).

41. Miyazawa, *Meiji ishin no saisōzō*, 17. Shimada's book is titled *Kaikoku shimatsu*. It appeared in English in 1896 H. Satoh as *Agitated Japan*.

42. Satoh and Shimada, *Agitated Japan*, 44.

43. Hill, "How to Write a Second Restoration," 342.

44. Marcus, *Paragons of the Ordinary*, 23-25.

45. Taguchi was born in 1855 to a shogunal official. He became a major figure in Meiji-era economic journalism and often wrote about history.

46. Kurimoto, *Hōan jisshū*, preface, 1-5.

47. Shimada, *Dōhōkai hōkoku* 1, no. 1, 17.

48. Ibid., 4, no. 2, 43.

49. Not all assessments of Oguri were positive on these points. Some accused him of being tricked by the French diplomat Léon Roches into building the Yokosuka shipyard, which would ultimately benefit the French more than Japan. See Emu, "Omoide no mama," 53. Others, such as Mishima Chūshū, imperial tutor to the Taishō emperor, disapproved of Oguri's hawkish stance against the Meiji forces (see Mishima, "Mishima Chūshū kō danwa," 86).

50. Togawa, "Shiden Oguri Kōzukenosuke," 30.

51. Ibid., 33.

52. Ibid., 37.

53. Togawa believed that one of Oguri's retainers was killed in his place. Supporters of Saigō Takamori also believed that their dead hero was living abroad and would soon return to Japan and achieve greatness (see Miyazawa, *Meiji ishin no*

saisōzō, 20); and Ravina, 174. The idea of a fallen, tragic hero who still lives goes back to the mid-Tokugawa period, when it was theorized that Yoshitsune had escaped to the Ezō region. For an analysis of the Yoshitsune legend, see Morimura, *Yoshitsune densetsu to Nihonjin*.

54. Togawa, "Oguri Kōzukenosuke," 37.

55. Ibid.

56. Akutsu, "Kurimoto Joun," 60. Oguri's former subordinate, Kurimoto Joun, likewise stated that he could not accept a position in the Meiji government even with the rationalization that he was working for the Meiji emperor. Moreover, Kurimoto avoided active participation in the Meiji government because of its refusal to acknowledge the achievements of the Tokugawa government (see Akutsu, 64–65).

57. One of Fukuzawa's closest students, Ishikawa Mikiaki, published the essay in 1901 in the newspaper *Jiji shinpō*. The title has been translated alternatively as "The Spirit of Manly Defiance," "Theory of Strained Endurance," and "Playing the Martyr." The title I use is by William Steele, whose translation, the only one in English, I reference here.

58. Itō, *Fukuzawa Yukichi no kenkyū*, 249.

59. Fukuzawa, *The Autobiography of Yukichi Fukuzawa*, 111–12.

60. Fujii, "Kaisetsu," in Kimura, *Sanjūnenshi*, 762.

61. Katsu, *Kaishū zadan*, 171. Kurimoto might have rubbed Yoshinobu the wrong way too. Kurimoto Joun stated that Yoshinobu relied completely on him to prepare the port opening in Hyōgo, breaking with tradition by allowing a retainer who did not have sufficient rank (Kurimoto was only a *metsuke*) to do the job. Kurimoto also related how Yoshinobu poured him some Western alcohol, presumably out of gratitude, during a visit to the Nijō Castle in Kyoto. Yoshinobu later denied all validity to Kurimoto's claims, saying that he only met Kurimoto once and did not give him Western alcohol, though he added that he often poured drinks for guests. See Tokugawa, *Sekimukai hikki*, 287–88.

62. Steele, "*Yasegaman no setsu*," 142.

63. Ibid., 145.

64. Matsumoto "kaisetsu" in Fukuzawa, *Fukuzawa Yukichi senshū*, 12:279.

65. Steele, "*Yasegaman no setsu*," 148. Katsu did not hold Fukuzawa in high opinion. He was asked about Fukuzawa during an interview, to which he responded, "Ah yeah, Fukuzawa, I know him. He came over once and never again. He became big, and that was it. He was the type of guy who liked to make money" (Itō, *Fukuzawa Yukichi no kenkyū*, 252).

66. Itō, *Fukuzawa Yukichi no kenkyū*, 255. It's also said that Tokutomi tried to steer people away from Fukuzawa's Keio juku school and into that of Niijima Jō.

67. Itō Masao, *Shiryō shūsei Meijijin no mita Fukuzawa Yukichi*, 33.

68. Ibid., 33–34.

69. For more on Tokutomi in English, see Pierson, *Tokutomi, 1863–1957*; Pyle, *The New Generation in Meiji Japan*; and Swale, "Tokutomi Sohō."

70. Pyle, 41.

71. Pierson, *Tokutomi Sohō, 1863–1957*, 262. Tokutomi's Restoration hero was Yoshida Shōin, whom he appropriated first as a revolutionary in his 1893 biography

but then as an exemplar of *kokutai*—a true patriot and imperial loyalist—in the 1908 revision.

72. Fukuzawa, *Fukuzawa Yukichi senshū*, 12:257.

73. Ibid., 12:259.

74. Itō, *Sabakuha no ketsujin*, 441.

75. For example, Katsu calls Kurimoto "petty" and accuses him of lying in his memoirs (*Kaishū zadan*, 170).

76. Katsu, *Hikawa seiwa*, 363.

77. Watanabe and Katsu, *Ishin genkun*.

78. Koizumi Takashi, "Kaisetsu," in Fukuzawa, *Meiji jūnen*, 134.

79. *Meiji jūnen*, 29.

80. Ibid., 45.

81. Lebra, *Above the Clouds*, 93. Later in his life, Fukuchi began work on a biography of Yoshinobu directed by Shibusawa Eiichi, but his brief time as a Diet member and subsequent illness forced him to drop out of the project.

82. Huffman, *Politics of the Meiji Press*, 127. I appreciate Jim Huffman pointing me to this anecdote.

83. Suzuki, "Fukuchi Ouchi no rekishikan," 18.

84. Shimane, *Tenkō*, 33.

85. Ibid., 185.

86. Huffman, *Politics of the Meiji Press*, 51.

87. Kazuo, "Bakufu suibōron wo yomu," 57.

88. Fukuchi, *Bakufu suibōron*, 147.

89. Ibid.

90. Fukuchi, *Bakumatsu seijika*.

91. Mizuno Tadanori was Fukuchi's mentor and sided with Oguri on the issue of fighting against the imperial forces. Fukuchi's gradualist approach and moderate political views are thought to derive from Mizuno. See Huffman, *Politics of the Meiji Press*, 77.

92. Fukuchi, *Bakumatsu seijika*, 254. Kurimoto and Oguri worked closely together, and Asahina served in the foreign, Edo, and finance commissioner offices.

93. Ibid., 274. *"Aa tendō ze ka hi ka."*

94. Ibid., 271.

95. Ibid., 273.

96. Ibid., 272.

97. Tsukagoshi founded the Kurabuchi Self Help Organization in 1889 (*Kurabuchi sonshi*, 1201). Tsukagoshi published other books during his time in Tokyo, such as biographies of Tokugawa Ieyasu and Tosa samurai Nonaka Kenzan.

98. Duus, "Whig History, Japanese Style," 435.

99. Ōkubo, *Nihon kindai shigaku no seiritsu*, 340.

100. Tsukagoshi, "Shiron: Oguri Jōshū," 441.

101. Translated in Carter, *Traditional Japanese Poetry*, 282.

102. Tsukagoshi, *Oguri Kōzukenosuke*, 545.

103. Ibid., 544.

104. Ibid., 545.

105. Ibid., 545.

106. Ibid., 90–91.

107. Seta, *Shōnen dokuhon #40*. Seta Tōyō, who also published under the name Seta Akiyuki, wrote several other books for Hakubunkan, including a similar, biographical reader about Konoe Tadahiro. Konoe, a member of the court close to the Shimazu family, was purged by Ii Naosuke.

108. Marcus, *Paragons of the Ordinary*, 24.

109. Yamaguchi, *"Haisha" no seishinshi*, 377.

110. Ibid., 3.

111. Ibid., 39.

112. Ibid., 65–66.

113. Ibid., 67–68. Satō Susumu was a military doctor and medical professor trained in Vienna who later became a peer. Asada Sōhaku, who practiced Chinese medicine, was the physician to the imperial court during the Meiji Restoration. It is possible that Seta is also commenting on the backwards ways of the imperial court.

114. Wigen, "Teaching about Home," 19.

115. Takahashi, *Kinsei Jōmō ijin den*, 102–4.

116. The *Usui County History* is one of the oldest local histories in Gunma. Although originally submitted to the Gunma Prefecture government in 1877, with two county histories, the Meiji edition is no longer extant. Usui and the other counties republished their histories during the Taishō period to celebrate the coronation of the Taishō emperor or to commemorate the dissolution of the counties. See Okada, "Gunma-ken ni okeru shishi hensan jigyō," 10.

117. Iwagami, *Shōgaku Kōzukeshi*, 50.

118. Ōtori knew Oguri through his French gunnery training, for which he received permission from Oguri himself. See Ōtsubo, *Oguri Kōzukenosuke kenkyū shiryō ochibohiroi*, 335.

119. Dansōsha, *Meika dansō* no. 14, 5.

120. Noguchi, "Karakkaze Akagisan," 172.

121. Nakajima, *Jōshū no Meiji ishin*, 194.

122. Ibid., 179.

123. For example, "bad mouthing" (*akugen*) against a Kantō inspector for halting local dance practice; see Ochiai, *Nihonshi liburetto*, 38.

124. It was mistakenly believed for many years that Ooto himself beheaded Oguri. See Abe, *Kaigun no senkusha Oguri Kōzukenosuke seiden*, 165.

125. Kodama, *Sugamo Diary*, 268.

126. *Ōtashi-shi tsūshihen kinsei*, 982–83.

127. Ibid., 984.

128. Shirayanagi, "Oguri Kōzukenosuke ibun," 28. Itō Chiyū was involved with the early Jiyūtō party and often heard such stories from Ooto's former colleagues.

129. Ochiai, *Nihonshi liburetto*, 91.

130. Imaizumi, *Essa sōsho*, 319.

131. *Takasaki shi shi* 5, 960–61.

132. See Nakajima Akira, *Jōshū no Meiji ishin*, 103. Hara Yasutarō verifies this during an interview with Abe Dōzan, in which he claims that Ooto gave him the money to build Oguri's grave. See Abe, *Kaigun no senkusha Oguri Kōzukenosuke seiden*, 172–73.

133. Hara Yasutarō, one of the vice inspectors who was a co-leader of the group sent to arrest Oguri, stated in an interview that "it is a fact that Oguri's head was stolen," although he was unsure of its final resting place. See Abe, *Kaigun no senkusha Oguri Kōzukenosuke seiden*, 174.

134. Typically, the name 誉田 is pronounced "Honda," but Nakajima's descendants state that the name is pronounced "Gonda," which is probably an association with their ancestor's village. See Koitabashi, "Oguri kankei no ayamari," 11–13.

135. Ikeda, "Tadamasako Gonda," 18. A document from the Annaka domain lists all of Oguri's items sent to the Tōsandō commander, including swords, rifles, armor, ammunition, clothing, crates, and jars. Some of these objects were sold for quick cash; others were kept by the soldiers. See *Annaka shishi*, 195–96.

136. See Yamakawa, *Aizu Boshin senshi*, 479–80. According to Aizu records, Tsukagoshi Tomiyoshi, Satō Ginjurō, and Satō Fukuyoshi were Oguri retainers who fought in Aizu. Oguri's wife is mentioned as having been escorted to Yokoyama Chikara's residence, but instead of Nakajima's daughter, Tsukagoshi Tomiyoshi's wife accompanied her. Also, a grave marker for at least one of these men, Satō Ginjurō, still stands in a former Aizu area.

137. Satō, "Watashi no ie," 25. Satō wandered throughout the area and died in Takasaki City in 1899.

138. Koitabashi, *Katsu Kaishū*, 120.

139. Ihara, *Hijitaka haku*, 403.

140. For more on the role of the fantastic, what Gerald Figal has referred to as the "*fushigi*" during late Tokugawa and early Meiji Japan, see chapter 1 of Figal, *Civilization and Monsters*.

141. Takasakishi, *Shinpen Kurabuchi sonshi 1, shiryōhen* 1, 624.

142. Kawaguchi, *Tokugawa maizōkin kenshō jiten*, 121. There is no evidence that Nakajima Kurando ever worked in the finance commissioner's office.

143. Inomata, *Kikigaki Inomata Kōzō jiden*, 181–82. Inomata, a socialist member of the lower house during the 1950s, was previously a lawyer on this case.

144. "Shinbun," *Yomiuri shinbun*, November 24, 1880, 1.

145. Taylor, "The Early Republic's Supernatural Economy," 8.

146. Foster, "Treasure Tales," 40–41.

147. Hunt, *Politics, Culture, and Class*, 40.

148. "Buried Treasure," 684.

149. Crossan, *Finding Is the First Act*, 32.

150. Ibid., 19.

151. Toyokuni, "Yokosuka kaikō," 3.

152. Kawaguchi, *Tokugawa maizōkin kenshō jiten*, 100.

153. Hatakeyama, *Maizōkin monogatari*, 1:19–23.

154. Locality is the key to understanding these legends. By Kawaguchi's count, twenty-seven out of the thirty-nine Tokugawa/Oguri buried-gold legends occur in Gunma Prefecture (Kawaguchi, *Tokugawa maizōkin kenshō jiten*, 98).

155. Akiyama, *Ōmiya zakkicho*, 2:200–201.

156. For an account of Tokutomi's various intellectual shifts, see Swale, "Tokutomi Sohō."

157. Miyazawa, "Bakumatsu ishin he no kaiki," 16.

158. Ibid., 14.
159. On this point about Saigō, see Ikai, "Shizoku hanran," 284.

Chapter Three

1. Fujitani, *Splendid Monarchy*.
2. See Karlin, "The Tricentennial Celebration." Tricentennial participants used much of the same commemorative grammar and vocabulary promoted by the Meiji government. For example, just as the government distributed pictures of the Meiji emperor and supporters cheered "Long live the emperor!" (*Tennō heika banzai*), Tokugawa supporters handed out pictures of Tokugawa Iesato, heir to the Tokugawa family, and cheered "Tokugawa Banzai." Organizers were careful to subordinate the Tokugawa shogunate to the imperial institution, such as when Enomoto Takeaki, leader of the pro-Tokugawa resistance in the north, noted Tokugawa Ieyasu's reverence for the emperor.
3. In the West, too, losers contested the mythical foundation of the modern state, as happened in France, where conservatives refused to participate in the national commemorations of the French Revolution but commemorated the births and deaths of the Bourbons (Gillis, ed., *Commemorations*, 8–9).
4. For more on Tokyo as a national space, see chapter 2 in Fujitani, *Splendid Monarchy*.
5. Abe, "Yokohama rekishi to iu rireki no shohō," 56. Yamato Takeru's statue is believed to be Japan's first bronze. It was erected in Kenrokuen Park to commemorate soldiers who died in Ishikawa Prefecture during the Restoration fighting there. Ōmura Masujirō's statue is in Yasukuni Shrine.
6. Fujitani, *Splendid Monarchy*, 124.
7. Ōkubo argues that he appears in casual garb because a popular audience liked this countrified version of Saigō, which made him seem more like them (Ōkubo, *Nihon kindai shigaku no seiritsu*, 430).
8. Bungei Shunjū, *Bungei shunjū ni miru Sakamoto Ryōma*.
9. Tokita, *Ze ya hi ya Ii Tairō*, 203.
10. Brownlee, *Japanese Historians*, 86.
11. Abe, "Yokohama rekishi to iu rireki no shohō," 49.
12. Ōkubo, *Sabakuha no rongi*, 95–96.
13. Abe Yasunari also argues in a separate article that the fiftieth anniversary of the opening of ports was conflated with the idea that Japan was open to the world for only fifty years. See Abe, "Kaikō gojūnen to Yokohama," 1–19.
14. Takada, "Ishin no kioku," 84.
15. Ibid., 77. Iwasaki, *Ishin zenshi*; *Sakurada gikyoroku*.
16. Ibid., 83.
17. Ōkuma's complete speech can be found in Tokita, *Ze ya hi ya Ii Tairō*, 211–22.
18. Lebra Takie Sugiyama, *Above the Clouds*, 48.
19. Ōkuma was not alone in reprising Restoration memory in a political speech. Years later, Hara Kei—from a Tokugawa loyalist domain and also a Satchō critic, noted his victory was a vindication for his province from the disgraceful reputation of "rebel army" (*zokugun*) (Lebra Takie Sugiyama, *Above the Clouds*, 92).

20. Ōkuma, *Fifty Years of New Japan*, 43 (originally published in Japanese in 1908). A former Hikone retainer, Nakamura Katsumaro, also published an Ii biography released in English in 1909; see Nakamura, *Lord Ii Naosuke and New Japan*.

21. Ōkuma, *Fifty Years of New Japan*, 83.

22. Miyake Setsurei in preface to Fujisawa, *Kakurō Andō Tsushima no kami*, 4.

23. For more on Ayako's relationship to Oguri Tadataka, and on the Ōkumas' trip to Tadataka's grave, see Ichishima, *Ōkuma kō ichigen ikkō*, 398–406. Apparently neither local people nor Ayako were completely aware of the grave, but one local man confirmed this for her and sent her a rubbing of the gravestone.

24. After the Restoration, Oguri's wife and newborn daughter eventually moved to Tokyo, where Minomura Rizaemon cared for them as he had promised Oguri years earlier. According to the financial accounts of the Minomura Gōmei company, the total amount of money listed under "Oguri Kuniko financial assistance" between 1877 to 1887 totals 1450 yen, 42 sen, and six rin. See Minomura, *Minomura Rizaemon den*. Yokiko, Mataichi's wife, eventually left the Oguri family after leaving Aizu with the rest of the Oguri women, and probably returned to her natal family in Tokyo. Mataichi's younger brother briefly served as the head of the Oguri family, taking the name Oguri Tadasachi, until Kuniko married Sadao.

25. Sadao's account of entering the Oguri family demonstrates how desperate people were to assist them. Sadao had been approached several times at the age of twenty-five and asked if he would become the heir to the Oguri family. He had refused because he had no plans on getting married until his mid-thirties, and he wanted to have a family of his own. When Oguri's wife died, leaving the daughter, Kuniko, on her own, Fujita Mokichi (journalist/politician) and Asabuki Eiji (Mitsui) suggested to Sadao that he marry her. He took their words under consideration, thinking that his own unhealthy constitution might be better served with a strong wife to take care of him. After meeting Kuniko at several social functions hosted by Ōkuma, however, he felt betrayed by his friends because of her small size and fragility. His friends assured him, "The Ōkuma family and Oguri relatives only want you to marry her to continue the Oguri name." In no way, they guaranteed, would he be responsible for any problems that might encumber the Oguri family—such problems would be taken care of by relatives. After much thought, Sadao agreed to the marriage, and the wedding took place at the Ōkuma residence in the spring of 1887. See *JOJ*, no. 187 (1932), 54–55.

26. Adachi, *Kaikoku shidan*, 406.

27. Ibid., 407.

28. Oguri Tadahito, "Nichirō sensō to Oguri Tadamasa," 12.

29. 仁義禮智信 (benevolence, morality, propriety, wisdom, loyalty). Oguri Sadao donated the vertical piece to the Fumon'in Temple in Ōmiya City; the other was given to the Tōzenji Temple long after WWII (ibid., 13).

30. Yokosukashi, *Yokosuka annaiki*, 116.

31. Ichikawa Yasō, "Oguri Jōshū kō," 29. These included documents sent from the Oguri family to the Tokugawa shogunate and a genealogy written by a local Gunma historian. Many of Oguri's possessions had been taken by the Tōsandō and sold. Due to his family needing to escape from the area, they took with them only what they could carry; thus, what little remained was at the Tōzenji or at the

Fumon'in in Saitama Prefecture. More significant items, such as Oguri's diary, had not been discovered by 1914. Other items at the Tōzenji, some of which may have been on display at Yokosuka, were lost during a temple fire.

32. Tsukagoshi, *Oguri Kōzukenosuke matsuro jiseki*, 1.

33. Ibid., 2.

34. Ibid., 3.

35. Ibid., 4.

36. Toyokuni, "Yokosuka kaikō gojūnen," 24.

37. Ibid., 26.

38. *JOJ* 65 (1922): 53. It is likely that Yokosuka already possessed adequate funds for the bust because the letter sent to *JOJ* was dated August, published in September, and the unveiling ceremony had been planned for September 27—the memorial day for the founding of the Yokosuka shipyard.

39. Ibid., 53.

40. Ibid. Opened in 1912, Suwa Park was close to the prewar Yokosuka City Hall and would have been the center of the city, suggesting that the busts were part of municipal, rather than naval, planning.

41. *JOJ* 67 (1922): 50.

42. Claudel, "Inauguration du buste de L. Verny," 160. Claudel has several long speeches about Yokosuka, Oguri, and Verny.

43. For an explanation into the origins of *zōi*, see Torao, "Zōi no shopoteki kōsatsu."

44. See Kondō's epilogue in Tajiri, *Zōi shoken den*, 874. Originally published in 1927, this is a collection of biographies of those who received court rank posthumously from 1868 to 1927. Kondō himself compiled biographies dating between 1927 to 1944, a time when, according to Kondō, promotion was sporadic. After the war, *zōi* was officially abolished, although Kondō claims it continued in other forms but without its previous "legal scope."

45. Maruyama, "Ogyū Sōrai no zōi mondai," 111–12. Some received several promotions; for example, Hirata Atsutane received a higher rank in 1943. As Maruyama notes, there were no explanations as to why some people were promoted ahead of others.

46. Takada, "Ishin no kioku," 76–77. Tanaka was famous for inventing the story that the Meiji empress saw Sakamoto Ryōma in a dream.

47. Shimane, *Tenkō*, 8.

48. Maruyama, "Ogyū Sōrai no zōi mondai," 116.

49. Ibid., 117. Original quote from Mikami Sanji, *Mainichi shinbun*, December 13, 1915. Mikami was a pro-emperor historian employed at the Tokyo Imperial University.

50. Walthall, *The Weak Body of a Useless Woman*, 248–50.

51. *Yomiura shinbun*, March 21, 1924, morning edition, 5.

52. Shimoda, "Between Homeland and Nation," 278. In 1876, the Dajōkan issued proclamation #108, which allowed all of those killed in the shogunate's military to be memorialized. Other vilified domains memorialized their warriors in the year immediately following the proclamation. See Takaki, "'Kyōdoai' to 'aikokushin' wo tsunagumono," 7.

53. Ibid., 292. It was not until 1965 that such rebels were enshrined in Yasukuni, and even then they were only enshrined in an annex.

54. Mehl, *History and the State*, 57–58.

55. Shidankai, *Senbō junnan shishi jinmeiroku*, prologue.

56. Ibid., 6–7.

57. Ibid., 58.

58. Toyokuni, "Yokosuka kaikō gojūnen shukuten ni saishite," 29.

59. He also erected a memorial stone to his fellow Shōgitai members in 1921.

60. Shimada was appointed in 1915. Oguri Sadao provided Shimada with documents to help the case (Shirayanagi, "Oguri Kōzukenosuke ibun," 30–31).

61. Toyokuni, "Yokosuka kaikō gojūnen shukuten ni saishite," 30.

62. *JOJ* 86 (1924): 2.

63. Although several other memos written to support Oguri's posthumous court rank also list him as minor fifth rank, I have seen no evidence for this before the twentieth century. Typically, current scholarship assumes that he received only the sixth rank in 1857 (see Ichikawa, Murakami, and Koitabashi, eds., *Oguri Kōzukenosuke*, 99). Indeed, even the author of this memo acknowledges that he is uncertain when Oguri received this rank, guessing 12/1858, when Oguri was promoted in his shogunate rank. Other candidates included Akimoto Okitomo, a viscount and member of the House of Peers, and Akimoto Yukitomo, the former daimyo of Tatebayashi.

64. Naimusho shohitsu, no. 93 (Japanese National Diet archives), 73.

65. Ibid., 74.

66. National Diet archives, Ondai 9, 105, 20 October 1928.

67. Ibid., 106.

68. See *JOJ* 140 (1928): 7–16.

69. Lebra Takie Sugiyama, *Above the Clouds*, 92.

70. *Yomiuri shinbun*, November 16, 1917, morning edition, 5.

71. Miyazawa, *Meiji ishin no saisōzō*, 87.

72. Ibid., 97.

73. Ōkubo, *Sabakuha no rongi*, 331.

74. Miyazawa, *Meiji ishin no saisōzō*, 99.

75. Ibid., 91–94.

76. Yoshinobu was finally willing to talk about the Restoration years after the Meiji emperor had awarded him several commendations and accolades, which included allowing Yoshinobu to establish his own family line. In 1907, the capitalist giant Shibusawa Eiichi initiated a project to interview and produce a biography about his former lord. Yoshinobu agreed only after Shibusawa promised not to publish anything until long after Yoshinobu's death. Thus began a series of interviews with Yoshinobu conducted by a team of historians under the title "Society for Dreams from the Past (*Sekimukai*)" (Kashima, "San-shimon shugisha Shibusawa Eiichi," 278–81). Yoshinobu's reminiscences, although an interesting resource for understanding the man himself, did not fundamentally change historians' understanding of the Restoration. "Dreams from the Past" seemed an apt description for the endeavor; many times he claimed not to remember certain events or did not know details, and his answers were often vague. When the interviewers asked

Yoshinobu if it was his idea to have the Shinsengumi join forces with Aizu soldiers, he simply said that it was not but he allowed them to go anyway (see Tokugawa, *Sekimukai hikki*, 177). He even relied on the writings of ex-retainers to respond to his interviewers. He had said almost nothing about Oguri, except to acknowledge his reform efforts within the shogunate.

77. See Miyazawa, "Bakumatsu ishin he no kaiki," 13–15.

78. Rinbara, "Shōwa shoki no bakumatsu 'monogatari,'" 461.

79. He was a distant Oguri relative through his mother's side (the son of Oguri's youngest sister-in-law), a high-profile member of the Red Cross, and a prolific Kyoto law scholar. During the Russo-Japanese war, he worked as a legal advisor and served in the Korean colonial government. He is probably most well-known for his defense of Japan's role in East Asia during the prewar period, especially Japan's intervention in China. His law scholarship includes the anti–pan-Asianism works *Kōshūwan no senryō to Karafuto no senryō* and *Ajia ni ikiru no michi*. For example, when Japan seized Kiaochow from Germany during WWI, many in Japan believed that returning leased and occupied territories would earn Japan praise in the international community; after all, some argued, fighting Germany was a justice issue, not a land grab. But Ninagawa passionately defended Japan's actions, citing Britain's possession of Gibraltar as evidence that wars were fought for the sake of power. Coox and Conroy, eds., *China and Japan*, 24. See also Ninagawa, *Kōshūwan no senryō*, and *Les Réclamations Japonaises et le droit international*, 34–35.

80. Oguri Tadahito, "Ninagawa hakase to sono meicho," 3–5. His books were well advertised in newspapers with national circulation.

81. Ibid., 2. Ninagawa first met Tanaka in France when they overlapped as students there. They shared lecture tours and Ninagawa's connections to the military. Conrad Totman said of Ninagawa's books, "They are so emotional, so imprecise, and so bitter that I found them quite unhelpful and ended up disregarding them" (Totman, *The Collapse of the Tokugawa Bakufu*, 562).

82. Brownlee, *Japanese Historians*, 113.

83. Akatsuka, "Tokushu: 'Oguri sama' no iru mura," 63. Ninagawa's Oguri biography went into its third printing ten days after it was published.

84. Robertson, "Les 'Bataillons Fertiles.'"

85. See, in order, Ninagawa, *Ishin zengo no seisō to Oguri Kōzuke no shi*, 141, 168, 171.

86. Ibid., 5.

87. Ibid., 175–76.

88. Ibid., 135–36. Ninagawa likewise belittled Katsu Kaishū, who he suggested was not a true samurai because three generations earlier his family bought their way into the samurai class (see ibid., 137).

89. Ibid., 194.

90. Ibid., 92.

91. Ibid., 55.

92. Ibid., 58.

93. Ibid., 20.

94. Ninagawa, prologue in *Ishin zengo no seisō to Oguri Kōzuke. Zoku*, 1.

95. Ibid., 323.

96. Ibid., 326.

97. See Jūbishi, *Oguri Kōzukenosuke no shi*, 127–28, for his praise of Ninagawa and his book. The latter third of Jūbishi's monograph, however, was authored by Ninagawa, who exhibits much of the same anti-Meiji, pro-Oguri sentiments he did in his own book.

98. Nakazato, *Dai bosatsu Tōge*.

99. Itō, *Sabakuha no ketsujin*, 435.

100. Ibid., 438.

101. *JOJ* 185 (1932): 51.

102. *JOJ* 139 (1928): 61.

103. Kanakura, *Furansu kōshi Rosesu to Oguri Kōzukenosuke*,142.

104. Ibid., 140. Kanakura argues that most of Oguri's ideas came from Roches. Although many believe that Oguri should be credited with importing the idea of a modern company to Japan, it was really Roches who should be called the father of the Japanese company (ibid., 387). Oguri's military strategy to attack the imperial forces was also probably Roches's idea (ibid., 398).

105. Ibid., 34.

106. Wigen, *A Malleable Map*, 191.

107. Vlastos, "Agrarianism without Tradition," 91–92.

108. Narita, *"Kokyō" to iu monogatari*, 55

109. Wigen, "Teaching about Home," 561.

110. Gunma Kyōikukai, ed., *Kyōdo dokuhon*, 1.

111. Ibid., 187–91.

112. Ichikawa had been active in the movement to have Oguri given posthumous honors.

113. Ichikawa Yasō, "Oguri Jōshū kō," 15. In Japanese, the first epitaph read, "*Bakumatsu no ijin Oguri Kōzukenosuke shūenchi*," and the second read, "*Ijin Oguri Kōzukenosuke tsumi nakushite kono tokoro ni kiraru.*"

114. This was not uncommon; people in Tokyo had been forced to first acquire approval in Tokyo before erecting a memorial stone to Kōga Gengo, a Tokugawa retainer who fought in Ezo against the imperial troops (see Shimane, *Tenkō*, 18–19). This story is famous among Oguri researchers and was retold most recently in the 2001 special edition of the magazine *Jōshūfu* that featured monuments in Gunma Prefecture. The story comes from Ichikawa Yasō, whose grandfather Motokichi encountered trouble with Takasaki police over the epitaph (Akatsuka, "Tokushu," 60–61).

115. Ninagawa and Abe, "Oguri Kōzukenosuke to Fumon'in," 20.

116. *Nihon shinbun*, May 12, 1932. Reprinted in *Tatsunami*, no. 1, 14. The stone washed away in a flood in 1935 and was recovered two years later, which prompted locals to create another group, The Association for the Preservation of Oguri Kōzukenosuke's Legacy. This institutionalized local commemoration of Oguri drew attention—and visitors—to Kurata and Ubuchi villages from throughout the Kantō area.

117. For example, Arai Nobushime wrote a 1934 article for *JOJ* detailing the escape of Oguri's family through Agatsuma County, Gunma (Arai, "Oguri Kōzukenosuke fujin to Agatsuma gun," 38–44). Another example is a record of Gonda Village

headman Satō Tōshichi, who traveled around the world with Oguri in 1860. See Arai, "Kenbei shisetsu Oguri Bingo no kami," 40–44.

118. For example, the earliest mention of Saitama people attending the Oguri-Verny memorial services in Yokosuka occurred in 1935.

119. Abe, *Kaigun no senkusha*, 34.

120. Ibid., 145–46.

121. Itō, "Ishinshi no kakinaoshi," 2.

122. Original in Nakazato, *Dai-bosatsu Tōge* 6, reprinted in *Ōmiyashi-shi* 4 (1982): 124.

123. Originally printed in *Saitama gōshikai*, May 1940; reprinted in *JOJ* 279 (1940).

124. Ibid., 8.

125. Ibid., 32. In the next sentence, he stated that she never once came to the Fumon'in either.

126. Ibid., 187–88.

127. Ibid., 181. According to Abe, Sadao wanted to give money to fix the graves, but this never materialized.

128. Another small town in Takasaki also claimed possession of Oguri's head, but its claims did not enter into the Tōzenji/Fumon'in competition, nor does any evidence exist to support its claims.

129. Hayakawa, "Bakumatsu no ijin," *JOJ* 70 (1923): 32–40.

130. Toyokuni, "Oguri Jōshū no shukyū," *JOJ* 220 (1935): 61–62.

131. See, for example, Hara's account in Mashimo, *Yanada sensekishi*, 222, and an account by a soldier stationed in Tatebayashi at the time (ibid., 479).

132. Abe, *Kaigun no senkusha*, 174. Another researcher asked Hara about the execution and reported that Hara could barely hear at all but stated, "he was killed by my hand" (*ore no te de kitta*), which was understood to mean that one of his men was ordered to behead Oguri. The logic here is that a man of Hara's rank would not behead a criminal. Instead, a local Annaka man admitted that his great uncle, Asada Gorō, executed Oguri. Hara had asked for volunteers to perform the execution and, having found no takers, ordered Asada, who was not only taller than the other Annaka domain samurai but a well-known swordsmen, to kill Oguri using Hara's sword.

133. Mashimo, *Yanada sensekishi*, 222.

134. Uchino, "Dōzan no Oguri Kōzukenosuke," 40.

135. Ibid., 48.

136. Yoshio Kodama, *Sugamo Diary*, 76. I could not find an exact date for Gotō's dig, but according to one website, he visited Mount Akagi in 1933.

137. Inomata and Tsuneo, *Kikigaki Inomata Kōzō jiden*, 185. Inomata notes that one of the Mizuno family was asked to testify about his family's connection to the buried-treasure legend after it appeared in an article during the late 1920s. The judge decided to give Seki a light prison sentence of only ten months, with three years' suspended sentence, because Seki honestly believed in the legend and was not intentionally trying to defraud people of their money.

138. See *Tokyo Asahi shinbun*, March 17, 1933, morning edition, 11; March 25, 1933, evening edition, 2; and February 2, 1935, evening edition, 2.

139. *JOJ* 294 (1941): 4.

140. Ibid., 7.

141. Abe, *Kaigun no senkusha*, 289.

142. *JOJ* 294 (1941): 7.

143. Abe, *Kaigun no senkusha*, 289. Abe noted with satisfaction that not long after their visit, the whole gang was arrested.

144. Abe pressed Sadao for a letter of introduction to Ozaki Yukiō, but when Sadao handed him Ozaki's business card, Abe demanded more. Sadao wrote to Abe asking him why he wanted an introduction to Ozaki in the first place: "Do you want to meet him because he is a big shot (*erai hito*) or do you want to get something out of him?" See Shirayanagi, "Oguri Kōzukenosuke ibun," 48–49.

145. See, for example, Sheldon Garon's study of the middle-class role in the creation of social management, *Molding Japanese Minds*; or for educational reform, see Platt, *Burning and Building*.

Chapter Four

1. Narita, "Historical Practice before the Dawn," 118.

2. Ibid., 119.

3. Gayle, *Marxist History*, 114. For Ishii Takashi's view, see Ishii, "Bakumatsu ni okeru han shokuminchika."

4. Gluck, "The People in History," 45.

5. Kitajima, ed., *Edo bakufu: Sono jitsuryokushatachi*, 269.

6. Oka, *Ii Tairō*. Oka had founded the San Francisco branch of the socialist group Commoner's Society (Heiminsha), to which Kōtoku Shūsui also belonged.

7. Miyazawa, *Meiji ishin no saisōzō*, 209–10.

8. Tamura, *Katsu Rintarō*, 16.

9. Ninagawa, *Ishin seikan*, 256–57.

10. Ibid., 35.

11. Ibid., 262.

12. Ibid., 261–62. These quotes come from Hayashi Razan, who was often thought to be one of the Tokugawa shogunate's official ideologues. Herman Ooms problematized this overemphasis on Hayashi's role in the shogunate. See Ooms, *Tokugawa Ideology*. Moreover, the word used for realm, *tenka*, is understood by Ninagawa as referring to the people, yet its true meaning was less democratic. During Oda Nobunaga's rule, *tenka* sometimes referred to Nobunaga himself (ibid., 33). In the Tokugawa period, *tenka* was understood as a realm to be ruled over by the shogun.

13. Ninagawa, *Ishin seikan*, 82. Yoshinobu wrote, "*hiroku tenka no kōgi wo tsukushi.*" The equivalent phrase in the Charter Oath is "*hiroku kaigo wo kesshi.*" Ninagawa's reading of Yoshinobu's memorial is, of course, extremely selective. Much of Yoshinobu's memorial is filled with pro-emperor sentiment.

14. "Dai tenkō wo togeta," *Yomiuri shinbun*, November 7, 1956. This article discusses Ninagawa's views of the Meiji Restoration and the emperor as discussed in his book, a best seller according to the article.

15. Ninagawa, *Tennō*, 146. Emperor Kaika is believed to be a legendary figure.

16. Ibid., 93.

17. Ibid., 172.

18. Ibid., 102.

19. Ninagawa, *Ishin seikan*, 270.

20. Ibid., 275.

21. Ninagawa, *Kaikoku no senkakusha*.

22. Tamamuro, *Saigō Takamori*, 1.

23. Ninagawa, *Kaikoku no senkakusha*, 89.

24. *Yomiuri shinbun*, July 11, 1956. It is also possible that Ninagawa feared repercussions for directly attacking the emperor system, which no longer concerned him after the war.

25. Gayle, *Marxist History*, 45. Unlike other leftist historians, Inoue shared with Ninagawa an optimistic view of the pre-Meiji history as being a democratic time, when people enjoyed peace and freedom.

26. Doak, "What Is a Nation and Who Belongs?" 304.

27. Gluck, "The Past in the Present," 80.

28. Gluck, "The 'End' of the Postwar," 293.

29. Gluck, "The People in History," 26. On the *minshūshi* movement of the 1960s and 1970s, see Fujitani, "*Minshūshi* as Critique," 303–22.

30. Shimazu, "Popular Representations of the Past," 103–5.

31. Tanaka, "Sakurada no yuki," 120. On the rewrite of "Sakurada mongai" to "Sakurada no yuki," see Ozaki, "Kaisetsu," 418.

32. Ozaki, "Kaisetsu," 441.

33. Kaneko, "Hana no shōgai wo miru," 63.

34. Suga, "Shinkokugeki," 51. Apparently Ono had wanted to write about other historical events, such as the 2/26 incident, to comment on contemporary society. His late decision to write about Ii is also a cause for the unevenness of the story. The opening day at Meijiza was delayed, the fourth time a play had been delayed in Meijiza's history, due to last minute script revisions. On the delay, see Hamada, "Ii tairō jōetsu chūshi," 42.

35. Kasahara, *Gendai ni ikiru Nihonshi no gunzō*, 242–44. Television adaptations of the book reappeared in 1974 and 1988.

36. Ibuse, *Ibuse Masuji jisen zenshu*, 3:401.

37. http://www7.ocn.ne.jp/~fumonin/sub2.htm. Contains pictures and introductions to Abe Dōzan's connection to Ibuse, Nakazato Kaizan, and Tokutomi Sohō.

38. Shinchosha, *Rekishi shōsetsu no seiki*, 783. A discussion of the story among Akiyama Shun, Katsumata Hiroshi, and Nawata Kazuo.

39. Liman, *Ibuse Masuji*, 250–51. See also Ibuse, *Castaways*, 9–10.

40. Treat, *Pools of Water*, 173.

41. Kume was a Meiji- and Taishō-period historian who participated in the Iwakura Mission of 1871 (Ibuse Masuji, "Oguri Kōzukenosuke shuzai dewa Gunma made"). A picture dated June 11, 1978, shows Ibuse and the Tōzenji Temple monk on Mount Kannon, where Oguri had been constructing a house. See *Tatsunami*, no. 4 (1979), inside cover.

42. Liman, *Ibuse Masuji*, 419.

43. Ibid.

44. Ibid., 250.

45. Ibuse, *Fumon'in san* in *Tanpen meisakusen*, 304.

46. Liman, *Ibuse Masuji*, 329.

47. Tanizaki, *Tanizaki Junichirō zenshū*, 18:409–10.

48. Ibuse, *Ibuse Masuji jisen zenshū*, 3:401.

49. The writing in the last version, to quote one critic, "is all over the place" (Shinchosha, *Rekishi shōsetsu no seiki*, 783).

50. My Town Saitama, July 4, 2007. http://mytown.asahi.com/saitama/news.php ?k_id=11000180704020001.

51. Hirano, *Mr. Smith Goes to Tokyo*, 66.

52. Tsutsui, *Jidaigeki eiga no shisō*, 51–53.

53. Hirano, *Mr. Smith Goes to Tokyo*, 83. The film was later produced by an independent company (1954).

54. Thornton, *The Japanese Period Film*, 38.

55. Satō, Tadao, *Iji no bigaku*, 120–21.

56. Ibid., 123.

57. *Samurai of the Great Earth* is based on Honjō Mutsuo's proletariat novel *Ishikarigawa*.

58. There is no extant copy of the 1956 version of this film, and I have relied on a widely used film database for the basic description. See http://www.kinejun.jp/cinema /%E9%8D%94%E9%B3%B4%E6%B5%AA%E4%BA%BA.

59. Ōsone, *Ōedo no kane: Dai Tokyo tanjō*. Although available in VHS format, most large movie rental outlets do not carry it, nor is it easy to find for purchase.

60. Fujita, *Eiga no naka no Nihonshi*, 135.

61. *Yomiuri shinbun*, August 6, 1958, 4.

62. Ibid.

63. Ibid.

64. Thornton, *The Japanese Period Film*, 66.

65. Desser, "Toward a Structural Analysis of the Postwar Samurai Film," 146.

66. Ibid., 147.

67. Saitō, "Sengo jidai shōsetsu no shisō," 18.

68. *Kinema junpō* 1035, no. 220:71.

69. Ibid., 71.

70. Saitō, "Sengo jidai shōsetsu no shisō," 13.

71. Narita, *Shiba Ryōtarō no bakumatsu Meiji*, 6. Shiba wrote over two hundred books in his lifetime. His collected works, including fiction, essays, and lectures, amount to sixty-eight volumes. Some of his work has also become the basis for movies and NHK historical television dramas. Donald Keene noted Shiba's ability to write in an exciting way and praised him for allowing the Japanese to be proud of their history at a time when Japanese history was "reduced to childish fantasy of costume movies." See Keene, *Five Modern Japanese Novelists*, 95.

72. Gluck, "The People in History," 26.

73. Keene, *Five Modern Japanese Novelists*, 90.

74. Shiba, *The Last Shogun*.

75. Narita, *Shiba Ryōtarō no bakumatsu Meiji*, 40.

76. Ibid., 73–75. Katsu, Fukuzawa, and Sakamoto each maintained a safe distance from the Meiji policy-making apparatus: Sakamoto through death; Katsu and Fukuzawa by being orbital yet influential thinkers.

77. Ibid., 69–70.

78. Saitō, "Sengo jidai shōsetsu no shisō," 99.

79. Ibid., 104. Gluck notes that Sakamoto and Katsu as they appear in Shiba's works were men in pursuit of prosperity, exemplifying the "optimism boom" of the 1960s (Gluck, "The Past in the Present," 75).

80. See, for example, Shiba, *Moeyoken*, 2:70, 80–81.

81. Ibid., 2:448.

82. Narita, *Shiba Ryōtarō no bakumatsu Meiji*, 121.

83. Kaionji, *Bakumatsu dōran no otokotachi* in *Kaionji Chōgorō zenshū*, 20:200–201. Saigō and Katsu were Kaionji's heroes because they were concerned about keeping the Western powers out of Japan's reform efforts. Kaionji is also a critic of Ninagawa's portrayal of Oguri.

84. Shiba, *Jūichibanme no shishi*, 148–49.

85. Shiba, *Shiba Ryōtarō rekishi kandan*, 352.

86. Jansen, *Sakamoto Ryōma and the Meiji Restoration*, preface, x.

87. The Oguri Memorial Society acknowledged this potential boom in tourism (Ichikawa Yasō, "Oguri Jōshū kō," 32).

88. Ikenami, *Sengoku to bakumatsu*, 179.

89. Ibid.

90. Robertson, *Native and Newcomer*, 33. Robertson argues that *furusato* dominated the national imagination during the post-postwar period.

91. I am borrowing Robertson's phrase; see ibid., 18.

92. See Ivy, *Discourses of the Vanishing*, chapter one, for more on the "Discover Japan" and "Exotic Japan" campaigns of the 1970s and 1980s. *Enka*, which also makes use of the *furusato* troupe, became a specific musical genre in the 1970s.

93. Ibid., 24.

94. *Yomiuri shinbun*, January 26, 1938, 7. Interestingly, the article notes that Oguri's "real" grave was located in Gonda Village. Sadao had created a new family grave in Tokyo. Mataichi authored a biography of his father, Sadao, and wrote a lengthy popular work about his famous grandfather that was not published in his lifetime. See Oguri Mataichi, *Ryūkei Yano Fumio-kun den*, 6.

95. Ichikawa Yasō, "Oguri Jōshū kō," 17.

96. *Yomiuri shinbun*, December 22, 2008. He was also faulted for having built the Yokosuka naval base (Nishikata, *Jōmō karuta no kokoro*, 127).

97. Nishikata, *Jōmō karuta no kokoro*, 110–11.

98. Ibid., 128–29. Another work claims that Kunisada, Takayama, and Omaeda Eigorō were the three men left out, but this work focuses on a sociological and educational use of the Jōmō cards compared to those found in other prefectures, and only briefly touches on the Jōmō origins. See Haraguchi, *Jōmō karuta*, 62.

99. Ichikawa Yasō, "Oguri Jōshū ko," 18. This society was created shortly before the fire that engulfed Tōzenji.

100. Ibid., 20.

101. Ibid., 21.

102. Ibid., 21.

103. Dower, "Peace and Democracy," 4.

104. Ichikawa Yasō, "Oguri Jōshū kō," 27.

105. Ikeda, "Oguri Kōzukenosuke to Katsu Kaishū," 17.

106. *Kurabuchi sonshi*, 546. Yokosuka sent thirty-five men from their aviation unit to Kawaura in May 1945 and another forty to Mizunuma in June.

107. Ichikawa Yasō, "Oguri Jōshū kō," 20. The original Yokosuka bust was melted down for its metal during the war and another had been put in its place. This replacement bust now rests in the Tōzenji courtyard.

108. Yoshida Nao, head of naval construction at Yokosuka, wanted to place Kurimoto's bust in Yokosuka alongside those of Oguri and Verny, but during negotiations, which were not proceeding well, Yoshida died. His family decided to give the bust to the Tōzenji. See Ōtsubo and Hozumi, *Oguri Kōzukenosuke*, 338–39.

109. *Shinpen Kurabuchi sonshi 2 shiryōhen* 2, 254.

110. Ichikawa Yasō, "Kōryū wo fukameru," 27.

111. These other sites include a park, the Karasu River dam, and a "health, recreation, reservation area" (ibid., 28). For more on the construction industry and its role in environmental destruction, see chapter 1 of McCormack, *The Emptiness of Japanese Affluence*. For more on the relationship between national government initiatives and support of local tourism and the leisure industry in general, see Leheny, *The Rules of Play*.

112. *Tatsunami*, no. 12 (1987): inside cover. The projected cost was 1.5 billion yen. See Takasakishi, *Shinpen Kurabuchi sonshi 2 shiryōhen* 2, 256.

113. On the Hamayū Sansō website, Oguri is clearly stated as being the connecting feature between Yokosuka and Kurabuchi. See http://www6.wind.ne.jp/hamayu /sisetu/sisetu.html.

114. Murakami Shōken kept a record of the number of visitors who signed their name in the visitors' log from 1970 to 1976 (an average of 795 visitors per year). He admits that many of those who visited asked questions about the buried-gold legend but was pleased that at least some people were seriously interested in Oguri's story. See Murakami, "Jōshū kō no ihin ni tsuite," 12.

115. Ibid., 13.

116. *Asahi shinbun*, January 28, 1952, 11.

117. Kuwata, *Nihon takarajima tanken*, 79.

118. Koitabashi, "Oguri no maizōkin," 11–14. Kawahara died in 1967, but his digging has been taken up by his son.

119. Oguri Kuniko, "Oguri Kōzukenosuke no mago kara," *Yomiuri shinbun*, September 16, 1956, 9.

120. Kiya, *Bakushin Oguri Kōzukenosuke*.

121. Kiya also communicated with Kaionji Chōgorō and at some point read Kaionji's 1942 historical fiction of Oguri. Kiya was impressed that Kaionji had such a high opinion of Oguri despite having been born in Kagoshima. Ibid., 14.

122. Ibid., 20.

123. Ibid., 17. Tsunabuchi also felt that Kiya was endangering his goal of elevating Oguri's image because the book belittled Katsu too much, such as when Kiya accused Katsu of just wanting "to raise his own position." The book used Katsu's active role in the final years of the shogunate to argue that Oguri also played an important role in the shogunate. Tsunabuchi also questioned the use of historical fiction as a way of writing a biography.

124. Brundage, ed., *Where These Memories Grow*, 9.

125. Kiya's contacts at Chūō Kōron had moved on, thus ending his chances of publishing with them. Tsunabuchi published his own biography of shogunal retainers in 1984, which included a short description of both Katsu and Oguri. See Tsunabuchi, *Bakushin retsuden*, 184–228.

Chapter Five

1. Morris-Suzuki, *The Past within Us*, 23.

2. For example, the media-induced fear over "compensated dating," in which teenage high school girls sell sex for brand-name goods. See Leheny, *Think Global, Fear Local*.

3. Narita, *"Sensō keiken" no sengoshi*, 248

4. Gluck, "The Invention of Edo."

5. *Shūkan Asahi*, March 13, 2009, 28.

6. Moriki, "Meiji Ishin rejiimu," 55.

7. *Asahi shinbun*, April 4, 2007, 34.

8. *Asahi shinbun*, December 6, 1997, 3.

9. *Shūkan shinchō*, no. 47, December 12, 1996, 154.

10. Ibid.

11. *Asahi shinbun*, April 15, 2007, 34.

12. Yamaki, "130 nen no onshū wo koete," 48–50.

13. Nume and Shōzō, "'Boshin sensō,'" 81.

14. Hoshi, *Yominaoshi Boshin sensō*, 191.

15. Ibid, 9.

16. "Aizuwakamatsushi shisei," 82.

17. Hoshi, *Yominaoshi Boshin sensō*, 93.

18. "Aizuwakamatsushi shisei," 83.

19. In fact, as Lebra notes, neither side of the Meiji Restoration monopolized the claim to be victims. Descendants of nobles, daimyo, and imperial loyalists all claim that their ancestors suffered from oppression. Lebra, *Above the Clouds*, 96.

20. *Mainichi shinbun seibu*, August 20, 2004, morning edition, 35.

21. Other reconciliations have been successful; for example, the mayors of Hikone and Mito cities attended the 150th anniversary of Ii's assassination in Tokyo in March 2010. These cities became friendship cities in 1968. The descendants of Kunisada Chūji, a Robin Hood–like figure active during the *bakumatsu* period in Gunma Prefecture, visited the graves of those killed by his minions with descendants of his victims. One descendant agreed to participate because he thought such public reconciliations were good for town development. Still, he said he held a grudge against Kunisada. *Mainichi shinbun*, June 1, 2007, 10.

22. "Aizuwakamatsushi shisei," 82.

23. Kawaguchi, *Tokugawa maizōkin kenshō jiten*, 120. The 1993 statistic was 21.7 percent (see *Mainichi shinbun Tokyo*, March 9, 1993, morning edition, 21).

24. Ibid.

25. From the TBS show *Shitteiru tsumori* by Nihon Terebi, May 2000. The documentary is about Oguri and the Tokugawa buried-gold legend. Fukuzawa stated that he had spent a total of one hundred million yen on his project.

26. This quote comes from the Shibukawa City, Akagi Village Commerce and Industry, http://www7.wind.ne.jp/akagi-shoko/maizo/maizo02.htm. The legend is part of local tourism.

27. Crossan, *Finding Is the First Act*, 30.

28. Mizuno, *Akagi ōgon tsuiseki*, 162. Emphasis my own.

29. Abe, *Atarerareruka inaka*.

30. Ibid.

· 31. Mizuno, *Akagi ōgon tsuiseki*, 189.

32. Yajima, *Oguri Kōzukenosuke Tadamasa*.

33. This is from a website by Toda Books, a local bookstore whose owner is an Oguri fan. He created a short bibliography of Oguri-related books, with comments by Murakami. http://www5.wind.ne.jp/cgv/oguri/book.htm.

34. The winner, who previously served as governor, renamed the Kōchi Airport after Sakamoto Ryōma. *Yomiuri shibun Seibu*, August 20, 2004, 35.

35. *Shūkan Asahi* 114 (4933), March 13, 2009, 28.

36. Handō, "Meiji ishin wa hijō no kaikaku datta," 128.

37. Sakamoto, *Bakumatsu ishin keizaijin*, 29.

38. Sakamoto, *Oguri Kōzukenosuke no shōgai*, 478.

39. Sakamoto, *Bakumatsu ishin keizaijin*, 204.

40. Shiba, *Meiji to iu kokka*, 38.

41. Narita, *Shiba Ryōtarō no bakumatsu Meiji*, 122.

42. Takahashi, *Nippon taihen*, 471.

43. Akatsuka, *Kimi wa Tomii poruka wo kiitaka*, 9.

44. Arai, "Nihon saisei wo saguru," 89.

45. Narumi, *Dotō sakamakumo*; Yoshioka, *Jipangu no fune*.

46. "Ima, Meiji ishin wo tō," 55.

47. It was republished in book form in 2002, 2006, and 2009. See Dōmon, *Shōsetsu Oguri Kōzukenosuke*.

48. Enomoto Aki "Kaisetsu" in *Shōsetsu Oguri Kōzukenosuke*, 661.

49. *Mainichi shinbun chihō Gunma*, 21 September 1999. Oguri appeared once before on an NHK history series called *Rival's Japanese History* (*raibaru nihonshi*). This thirty-minute episode broadcast in 1995 pits Oguri against Katsu.

50. See for example Maehara, "Medeia sangyō to kankō sangyō."

51. *Jōmō shinbun*, November 30, 2000, 10–11. The French scholar published a book regarding the French support and construction of Yokosuka.

52. Ibid., 11.

53. Akatsuka, "Tokushu: 'Oguri sama,'" 65.

54. *AERA*, January 5, 1994, 47. Every Valentine's Day during the mid-1990s, young women were leaving chocolates, flowers, and love notes at Hijikata Toshizō's grave.

55. Matsuoka Masao, *Sōmu iinkai 8* (159), March 30, 2004.

56. Ibid.

57. Nakamura and Kishō, "Shōgatsu jidaigeki," 25.

58. On his temple website, Murakami Taiken invites visitors to join "the group to remove Kanrin maru from the history books." Katsu Kaishū was onboard the Kanrin maru, which anti-Katsu people believe is the only reason why the Kanrin Maru, and not Oguri's Powhatan, appears in textbooks. http://tozenzi.cside.com/kanrinmaru -byou.htm.

59. Summary of the post-cabinet meeting press conference with the minister of justice, February 8, 2008. Accessible on the Ministry of Justice website, http://www .moj.go.jp/hisho/kouhou/kaiken_point_sp080208-01.html.

60. Yamauchi, *Bakumatsu ishin ni manabu genzai*, 46.

61. Kimura, *Tengai no bushi*, vol. 4, 252.

62. Personal communication between Kimura and author, 6 August 2009.

63. Kimura, *Tengai no bushi*, vol. 1, 151.

64. Ibid., 189.

65. Ibid., 190.

66. Kimura, *Tengai no bushi*, vol. 2, 40.

67. Ibid., 55.

68. Ibid., 54.

69. Kimura, *Tengai no bushi*, vol. 4, 230–34.

70. Personal communication.

71. Kimura, *Tengai no bushi*, vol. 1, 205.

72. Ibid.

73. Ibid.

74. Kimura, *Tengai no bushi*, vol. 2, 201.

75. Ibid., 134.

76. Ibid., 201.

77. Kimura, *Tengai no bushi*, vol. 3, 255.

78. Ibid.

79. Kimura, *Tengai no bushi*, vol. 4, 255.

80. Nihon Kanji Nōryoku Kentei Kyōkai, http://www.kanken.or.jp/kanji/kanji 2007/kanji.html.

81. Kimura, *Tengai no bushi*, vol. 4, 255.

82. Ibid.

83. Personal communication.

84. Yokosuka City transferred full ownership of the Hamayū hot spring resort to Kurabuchi in 2005.

85. Ertl, "Revisiting Village Japan," 192.

86. "Oguri no sato seibi kihon keiaku," http://www.city.takasaki.gunma.jp/shisho /kurabuchi/chiiki/oguri.htm.

87. Although Aizu samurai were sent to the Yokosuka area for costal defense before Oguri built Yokosuka, the Oguri connection has been highlighted as the primary historical connection. See National Diet record 164, "gyosei kaikaku ni kansuru tokubetsu iinkai" #13, April 19, 2006.

88. Yamazaki, *Takai Kōzan yumemonogatari*.

89. *Asahi shinbun*, April 2, 2008, 12.

90. *Tatsunami*, no. 33 (2008): 21–23.

91. Ichikawa and Murakami, *Bakumatsu kaimei no hito*, 67.

92. *Kōtō gakkō Nihonshi A kaiteiban shidō to kenkyū*, 289.

93. *Kōtō gakkō Nihonshi A*, 138.

94. *Kōtō gakkō Nihonshi A kaiteiban shidō to kenkyū*, 290.

Conclusion

1. Regarding his writings on the Shinsengumi, Shiba stated that he just could not shake the influence of Shimozawa Kan's prewar study of the group. Tsuzuki-dani, "Shinsengumi 'fukken' he no keifu," 14.

2. *Miyazawa, Meiji ishin no saisōzō*, 136.

3. Haraguchi and Yukio, "Gōshi karuta asobi to gōshi ninshiki no keisei," 40–41.

4. "Yūsei mineika ni kansuru tokubetsu iinakai," no. 15, August 15, 2005; and Sakaya Taichi, "Sangiin kokumin seikatsu-keizai ni kansuru chōsakai," no. 1, 28 January 2009.

5. "Tokugawa maizōkin 'ana hori asobi' ni yonoku en shusshi shite kaisha shachō," 38.

6. *Mainichi shinbun Tokyo*, January 18, 2000, 24. See also *Mainichi shinbun chihōban Yamanashi*, February 16, 2000.

7. http://tozenzi.cside.com/.

8. "Meiji no chichi: Oguri Kōzukenosuke," 6.

9. Exhibit explanation on Tōzenji website, http://tozenzi.cside.com/aroundw -newyork.html.

10. *Oguri Kōzukenosuke jōhō*, December 2009.

11. Karatani, *Origins of Modern Japanese Literature*, 34.

12. "Yamaguchiken Hagishi kara," *Yomiuri shinbun Fukushima*, 3 March 2011, 17.

13. "Byakkotai jijin no chi nado Yamaguchiken Hagi shichora hōmon," *Yomiuri shinbun Fukushima*, October 28, 2011, 28. This White Tiger Brigade was a reserve unit of teenage samurai, some of whom committed suicide. Others fought in the Boshin War.

14. Takahori, "'Hisaichi Fukushima' kara," 3.

15. Ibid., 1

16. Ibid., 3.

Bibliography

Abe Dōzan. *Kaigun no senkusha Oguri Kōzukenosuke seiden*. Tokyo: Kaigun Yūshūkai, 1941.

Abe Issei. *Atarerareruka inaka: Tokugawa Maizōkin 120 Nenme No Chōsen*. Documentary. Wireworks, 2006.

Abe Yasunari. "Kaikō gojūnen to Yokohama no rekishi hensan." *Hitotsubashi Ronsō* 117, no. 2 (1997).

———. "Yokohama rekishi to iu rireki no shohō." In *Kioku no katachi: komemoreishon no bunkashi*. Edited by Abe Yasunari [et. al.]. Tokyo: Kashiwa Shobō, 1999.

Adachi Ritsuen. *Kaikoku shidan*. Tokyo: Chugai Shogyō Shinpō Shokyōsha, 1905.

"Aizuwakamatsushi shisei hyakushūnen kinen jigyō." *Tōhoku zaikai* 25, no. 6 (1999): 80–83.

Akatsuka Yukio. *Kimi wa Tomii poruka wo kiitaka: Oguri Kōzukenosuke to Tateishi Onojirō no "bakumatsu"* Nagoya: Fūbaisha, 1999.

———. "Tokushu: 'Oguri sama' no iru mura." *Jōshūfu* 8 (2001): 56–77.

Akimoto Shunkichi. *Lord Ii Naosuke and New Japan*. Tokyo: Japan Times, 1909.

Akiyama Kikuo. *Ōmiya zakkicho 2, Maboroshi no tera*. Ōmiya: Maruoka Shoten, 1974.

Akutsu Suguru. "Kurimoto Joun: A shogunal reformer in late Tokugawa Period." Master's thesis, Washington University, 1971.

Anderson, Joseph L, and Donald Richie. *The Japanese Film: Art and Industry*. Princeton, NJ: Princeton University Press, 1982.

Annaka shishi. Annaka-shi: Annaka-shi shi hensan iinkai, 1964.

Arai Kimio. "Nihon saisei wo saguru intabyuu shiriizu: Arai Kimio Tōkyū Eejienshii shacho "senkenryoku to danshiki" no Oguri Tadamasa kei riidaa ga shinriteki fukyō wo daikai suru." *Sapio* 11, 225 (1999): 87–89.

Arai Nobushime. "Kenbei shisetsu Oguri Bingo no kami no zuikōin Satō Tōshichi no sekai isshūki wo hakken." *Jōmō oyobi Jōmōjin*, 220 (1934): 40–44.

———. "Oguri Kōzukenosuke fujin to Agatsuma gun: sono kūshin tōsō no seki." *Jōmō oyobi Jōmōjin*, 211 (1934): 38–44.

Assmann, Jan, and John Czaplicka. "Collective Memory and Cultural Identity." *New German Critique*, no. 65 (April 1, 1995): 125–33.

Banno Junji. *The Establishment of the Japanese Constitutional System*. Nissan Institute: Routledge Japanese Studies Series, 1995.

Beasley, W. G. *Historians of China and Japan*. London: Oxford University Press, 1961.

Berlinguez-Kōno, Noriko. "How did Saigō Takamori Become a National Hero after His Death? The Political Uses of Saigō's Figure and the Interpretation of Seikanron." In *The Power of Memory in Modern Japan*, edited by Sven Saaler and Wolfgang Schwentker. Folkestone, UK: Global Oriental, 2008.

Bodnar John. "Pierre Nora, National Memory, and Democracy: A Review." *Journal of American History* 87, no. 3 (2000): 951–63.

———. *Remaking America: Public Memory, Commemoration, and Patriotism in the Twentieth Century*. Princeton, NJ: Princeton University Press, 1993.

Botsman, Daniel V. *Punishment and Power in the Making of Modern Japan*. Princeton, NJ: Princeton University Press, 2005.

Broadbridge, Seymour. "Shipbuilding and the State in Japan since the 1850s." *Modern Asian Studies* 11, no. 4 (1977): 601–13.

Brownlee, John S. *Japanese Historians and the National Myths, 1600–1945: The Age of the Gods and Emperor Jinmu*. Tokyo: University of Tokyo Press, 1997.

Brundage, Fitzgereald. *Where These Memories Grow: History, Memory, and Southern Identity*. Chapel Hill: University of North Carolina Press, 2000.

Bungei Shunjū Kabushiki Kaisha. *Bungei shunjū ni miru Sakamoto Ryōma to bakumatsu ishin*. Tokyo: Bungei Shunjū, 2010.

"Buried Treasure." *Harpers Young People*. August 22, 1882.

"Byakkotai jijin no chi nado Yamaguchiken Hagi shichora hōmon." *Yomiuri shinbun Fukushima*, October 28, 2011.

Calman, Donald. *The Nature and Origins of Japanese Imperialism: A Reinterpretation of the Great Crisis of 1873*. New York: Routledge, 1992.

Carter, Steven D. *Traditional Japanese Poetry: An Anthology*. Stanford: Stanford University Press, 1991.

Certeau, Michel de. *The Writing of History*. New York: Columbia University Press, 1988.

Clark, Edward Warren. *Katz Awa, "the Bismarck of Japan"; or, The Story of a Noble Life*. B. F. Buck, 1904.

Claudel, Paul. "Inauguration du buste de L. Verny." In *Cahiers Paul Claudel 1*. Paris: Gallimard, 1959.

Confino, Alon. "Collective Memory and Cultural History: Problems of Method." *American Historical Review* 102, no. 5 (1997): 1386–1403.

———. *The Nation as a Local Metaphor: Württemberg, Imperial Germany and National Memory, 1871–1918*. Chapel Hill: University of North Carolina Press, 1997.

Connerton, Paul. *How Societies Remember*. Cambridge: Cambridge University Press, 1989.

Coox, Alvin D., and Hilary Conroy, eds. *China and Japan: Search for Balance since World War I*. Santa Barbara: ABC-Clio Books, 1978.

Crane, Susan A. "Writing the Individual Back into Collective Memory." *American Historical Review* 102, no. 5 (1997): 1372–85.

Crossan, John Dominic. *Finding Is the First Act: Trove Folktales and Jesus' Treasure Parable*. Philadelphia: Fortress Press, 1979.

Dai Nihon ishin shiryō kōhon (DNISK). Historiographical Institute, University of Tokyo.

Dajōkan, ed. *Fukkoki*. Tokyo: Nagai Shoseki, 1929–1931.

Dansōsha. *Meika dansō*. Tokyo: Dansōsha, 1895.

Davis, Sandra T. W. "Ono Azusa and the Political Change of 1881." *Monumenta Nipponica* 25, no. ½ (1970): 137–54.

Desser, David. "Toward a Structural Analysis of the Postwar Samurai Film." In *Reframing Japanese Cinema: Authorship, Genre, History*, edited by Arthur Nolletti Jr. and David Desser, 145–64. Bloomington: Indiana University Press, 1992.

Doak, Kevin M. "What Is a Nation and Who Belongs? National Narratives and the Ethnic Imagination in Twentieth-Century Japan." *American Historical Review* 102, no. 2 (1997): 283–309.

Dōmon Fuyuji. *Bakumatsu Nihon no keizai kakumei*. Tokyo: TBS Buritanika, 1990.

———. *Shōsetsu Oguri Kōzukenosuke: Nihon no kindaika wo shikaketa otoko*. Tokyo: Shūeisha, 2006.

Dower, John. "Peace and Democracy in Two Systems: External Policy and Internal Conflict." In *Postwar Japan as History*, edited by Andrew Gordon, 3–33. Berkeley: University of California Press, 1993.

Duus, Peter. "Whig History, Japanese Style: The Min'yusha Historians and the Meiji Restoration." *Journal of Asian Studies* 33, no. 3 (1974): 415–36.

Eidson, John. "Which Past for Whom? Local Memory in a German Community during the Era of Nation Building." *Ethos* 28, no. 4 (2000): 575–607.

Enomoto Aki. "Kaisetsu." In *Shura wo iki himei ni shisu: Shōsetsu Oguri Kōzukenosuke Tadamasa*. Tokyo: Shueisha, 2010.

Ericson, Mark D. "The Bakufu Looks Abroad: The 1865 Mission to France." *Monumenta Nipponica* 34, no. 4 (1979): 383–407.

———. "Oguri Tadamasa: A Political Biography." Master's thesis, University of Hawaii, 1972.

Ertl, John Josef. "Revisiting Village Japan." PhD diss., University of California–Berkeley, 2007.

Figal, Gerald A. *Civilization and Monsters: Spirits of Modernity in Meiji Japan*. Durham, NC: Duke University Press, 1999.

Fine, Gary Alan. *Difficult Reputations: Collective Memories of the Evil, Inept, and Controversial*. Chicago: University of Chicago Press, 2001.

Foster, George M. "Treasure Tales, and the Image of the Static Economy in a Mexican Peasant Community." *Journal of American Folklore* 77, no. 303 (1964): 39–44.

Fraser, Andrew. "The Expulsion of Ōkuma from the Government in 1881." *Journal of Asian Studies* 26, no. 2 (1967): 213–36.

Freeman, Laurie Anne. *Closing the Shop: Information Cartels and Japan's Mass Media*. Princeton, NJ: Princeton University Press, 2000.

Frost, Peter K. *The Bakumatsu Currency Crisis*. Harvard East Asian Monographs 36. Cambridge: Harvard University Press, 1970.

Fujii Sadafumi, "Kaisetsu." In *Sanjūnenshi*. Tokyo: Tokyo Daigaku Shuppankai, 1978.

Fujisawa Morihiko. *Kakuro Andō Tsushima no kami*. Tokyo: Yūrindō Shōku, 1914.

Fujita Masayuki. *Eiga no naka no Nihonshi*, 1997. Tokyo: Chirekisha, 1997.

Fujitani Takashi. "*Minshūshi* as Critique of Orientalist Knowledges." *Positions* 6, no. 2 (1998): 323–44.

——. *Splendid Monarchy: Power and Pageantry in Modern Japan.* Berkeley: University of California Press, 1996.

Fukuchi Gen'ichirō. *Bakufu suibōron.* 1892. Tokyo: Heibonsha, 1967.

——. 1900. *Bakumatsu seijika.* Tokyo: Heibonsha, 1989.

Fukui Nakako. "Kōkei no chigyōjo no shihai: Todamin to Imagawamin." In *Kinsei kokka no kenryoku kōzō: seiji, shihai, gyōsei,* edited by Ōishi Manabu, 295–346. Tokyo: Iwata Shoin, 2003.

Fukuzawa Yukichi. *Commentary on the National Problems of 1877,* n.d.

——. *Fukuzawa Yukichi senshū.* Tokyo: Iwanami Shoten, 1980.

——. *Meiji jūnen teichū kōron, yasegaman no setsu.* Tokyo: Kodansha, 1985.

Fukuzawa Yukichi, and Eiichi Kiyooka. *The Autobiography of Yukichi Fukuzawa.* New York: Columbia University Press, 2007.

"Fumon'in." http://www7.ocn.ne.jp/~fumonin/sub2.htm.

"'Fumon'in san,' Ibuse Masuji." http://mytown.asahi.com/saitama/news.php?k_id=11000180704020001.

Gaku Shin'ya. *Shura wo iki himei ni shisu: Shōsetsu Oguri Kōzukenosuke Tadamasa.* Tokyo: Shueisha, 2010.

Garon, Sheldon M. *Molding Japanese Minds: The State in Everyday Life.* Princeton, NJ: Princeton University Press, 1997.

Gayle, Curtis Anderson. *Marxist History and Postwar Japanese Nationalism.* New York: Routledge Curzon, 2003.

——. "Progressive Representations of the Nation: Early Post-War Japan and Beyond." *Social Science Japan Journal* 4, no. 1 (2001): 1–19.

Gerson, S. "Une France Locale: The Local Past in Recent French Scholarship." *French Historical Studies* 26, no. 3 (2003): 539–59.

Gildea, Robert. *The Past in French History.* New Haven: Yale University Press, 1996.

Gillis, John R., ed. *Commemorations: The Politics of National Identity.* Princeton, NJ: Princeton University Press, 1996.

Gluck, Carol. "The 'End' of the Postwar: Japan at the Turn of the Millennium." *Public Culture* 10, no. 1 (1997): 1–23.

——. "The Invention of Edo." In *Mirror of Modernity: Invented Traditions of Modern Japan,* edited by Stephen Vlastos, 262–84. Berkeley: University of California Press, 1998.

——. "The Past in the Present." In *Postwar Japan as History,* edited by Andrew Gordon, 64–95. Berkeley: University of California Press, 1993.

——. "The People in History: Recent Trends in Japanese Historiography." *Journal of Asian Studies* 38, no. 1 (1978): 25–50.

Gordon, Andrew. *Postwar Japan as History.* Berkeley: University of California Press, 1993.

Gunma kenshi Hensan Iinkai. *Gunma kenshi. 20, shiryōhen. 10, kinsei 2, seimo chiiki. 2.* Maebashi: Gunma-ken, 1978.

Gunma Kyōikukai, ed. *Kyōdo dokuhon.* Maebashi: Kankodō, 1941.

Gunmaken bunka jigyō shinkōkai, ed. *Gunmaken shiryōshū. 5.* Maebashi: Gunmaken Bunka Jigyō Shinkōkai, 1969.

———. *Gunmaken shiryōshū.* 7. 1972.

Gyōsei kaikaku ni kansuru tokubetsu iinkai 13. National Diet Record, April 19, 2006.

Hamada Yūjirō, "Ii tairō jōetsu chūshi." *Kabuki tenbō* 2, no. 11 (1952): 42–48.

Hanashiro, Roy S. *Thomas William Kinder and the Japanese Imperial Mint, 1868–1875.* Boston: Brill, 1999.

Handō Kazutoshi. "Meiji ishin wa hijō no kaikaku datta: bakumatsusi to heisei Nihon ano toki mo gaiatsu to fukyō ga." *Bungei Junshū* 87, no. 6 (2009): 122–34.

Hanes, Jeffrey E. *Image and Identity: Rethinking Japanese Cultural History.* Kobe: Research Institute for Economics and Business Administration, Kobe University, 2004.

Haraguchi Mikiko. *Jōmō karuta: Sono Nihon hitotsu no himitsu.* Maebashi: Jōmō Shinbunsha, 1996.

Haraguchi Mikiko, and Yukio Yamaguchi. "Gōshi karuta asobi to gōshi ninshiki no keisei: Gunmaken no 'Jōmō karuta' no baai." *Gunma daigaku jissen kenkyū,* no. 11. Gunma Karuta, Jōmō Karuta Kenkyū Ronshū (1994): 1–44.

Hashimoto Takehiko. "Introducing a French Technological System: The Origin and Early History of the Yokosuka Dockyard." *East Asian Science, Technology, and Medicine* 16 (January 1, 1999): 53–72.

Hatakeyama Kiyoyuki. *Maizōkin monogatari: Ruporutaju.* 1. Tokyo: Jinbutsu Ōraisha, 1961.

Hayakawa Keison. "Bakumatsu no ijin Oguri Kōzukenosuke 2." *Jōmō oyobi Jōmōjin,* 70 (1923): 32–39.

Hearn, Lafcadio, Oliver Wendell Holmes Collection (Library of Congress), and Pforzheimer Bruce Rogers Collection (Library of Congress). *Out of the East: Reveries and Studies in New Japan.* Cambridge: Riverside Press, 1895.

Hill, Christopher. "How to Write a Second Restoration: The Political Novel and Meiji Historiography." *Journal of Japanese Studies* 33, no. 2 (2007): 337–56.

Hirano, Kyoko. *Mr. Smith Goes to Tokyo: The Japanese Cinema under the American Occupation, 1945–1952.* Smithsonian Studies in the History of Film and Television. Washington, DC: Smithsonian Institution, 1992.

Honjō Mutsuo. *Ishikarigawa.* Tokyo: Daikandō, 1939.

Hoshi Ryōichi. *Jōshū Gonda-mura no shūu: Oguri Kōzukenosuke no shōgai.* Tokyo: Kyōiku Shoseki, 1995.

———. *Saigo no bakushin Oguri Kōzukenosuke.* Tokyo: Chūkō Bunko, 2000.

———. *Yominaoshi Boshin sensō: Bakumatsu no tōzai tairitsu.* Tokyo: Chikuma Shobō, 2001.

Howell, David L. "Hard Times in the Kantō: Economic Change and Village Life in Late Tokugawa Japan." *Modern Asian Studies* 23, no. 2 (1989): 349–71.

Huffman, James L. *Creating a Public: People and Press in Meiji Japan.* Honolulu: University of Hawai'i Press, 1997.

———. *Politics of the Meiji Press: The Life of Fukuchi Genichiro.* Honolulu: University Press of Hawaii, 1980.

Hunt, Lynn. *Politics, Culture, and Class in the French Revolution.* Studies on the History of Society and Culture. Berkeley: University of California Press, 1984.

Ibuse Masuji. *Castaways: Two Short Novels.* Translated by Anthony Liman. Tokyo: Kodansha International, 1993.

———. *Fumon'in san*. In *Tanpen meisakusen 1925–1949: bushitachi no jidai*. Edited by Hirabayashi Fumio. Tokyo: Kasama Shoin, 1999.

———. *Ibuse Masuji jisen zenshū*. Vol. 3. Tokyo: Chikuma Shobō, 1996.

———. "Oguri Kōzukenosuke shuzai deha Gunma made." Circa 1978. In the possession of the Tōzenji archives.

Ichikawa Koichi, and Murakami Taiken. *Bakumatsu kaimei no hito Oguri Kōzukenosuke*. Takasaki: Gunmaken Takasaki Zaimu Jimusho, 1994.

Ichikawa Koichi, Murakami Taiken, and Koitabashi Ryōhei, eds. *Oguri Kōzukenosuke*. Maebashi: Miyama Bunko, 2004.

Ichikawa Yasō. "Kōryū wo fukameru Yokosukashi to Kurabuchimura." *Tatsunami*, no. 7 (1982): 24–31.

———. "Oguri Jōshū kō 'kenshō no ayumi.'" *Tatsunami*, no. 1 (1976): 1–32.

Ichishima Kenkichi. *Ōkuma kō ichigen ikkō*. Tokyo: Waseda Daigaku Shuppanbu, 1922.

Ienaga Saburo kyoju Tokyo kyōiku daigaku taikan kinen ronshu kanko iinkai. *Kindai Nihon no kokka to shisō*. Tokyo: Sanseidō, 1979.

Igarashi Eikichi, Hideomi Takahashi, and Shigenobu Ōkuma. *The National Wealth of Japan*. National Wealth of Japan Pub. Office, 1906.

Ihara Ryūjirō. *Hijikata haku*. Tokyo: Ihara Ryūjirō, 1913.

Ikai Takaaki. *Saigō Takamori: Seinan sensō he no michi*. Tokyo: Iwanami Shoten, 1992.

———. "Shizoku hanran to Saigō densetsu." In *Meiji ishin to bunmei kaika, Nihon no jidaishi 21*. Edited by Matsuo Masahito, 275–302. Tokyo: Yoshikawa Kōbunkan, 2004.

Ikeda Sazen. "Oguri Kōzukenosuke to Katsu Kaishū." *Tatsunami*, no. 3 (1978): 17–23.

———. "Tadamasako Gonda Insei Yowa." *Tatsunami*, no. 8 (1983): 15–18.

Ikenami Shōtarō. *Sengoku to bakumatsu: Ransei no otokotachi*. Tokyo: Kadokawa Shoten, 1980.

Ikoma Chūichirō. *Tokugawa bakufu no maizōkin: Oguri Kōzukenosuke Tadamasa no shōgai*. Nagoya: KTC Chūō Shuppan, 1998.

"Ima, Meiji ishin wo tō." *Kan*, no. 13 (2003): 29–297.

Imaizumi Takujiro. *Essa sōsho*. Nagaokas: Essa Sōsho Kankōkai, 1900.

Inomata Kōzō, and Yamashita Tsuneo. *Kikigaki Inomata Kōzō jiden: Ichi musantō bengoshi no Shōwa shi*. Tōkyō: Shisō no Kagakusha, 1982.

Inoue Masanao. "Inoue Masanaokun kyūjidan." *Kyūbakufu* 5, no. 4 (1901): 33–45.

Ishihara Kaizō. *Bakumatsu no eiketsu Oguri Kōzukenosuke wo shinobu: Yokosuka kaigun kōshō sōsetsu no yurai*. Tokyo: Manejimentosha, 1934.

Ishii Takashi. *Bakumatsu hiun no hitobito*. Yokohama: Yūrindō, 1979.

———. "Bakumatsu ni okeru han shokuminchika no kiki to tōsō: 2." *Rekishi Hyōron*, no. 33 (1951): 1–16.

———. *Meiji ishin no kokusaiteki kankyō*. Tokyo: Yoshikawa Kōbunkan, 1966.

Itō Chiyū. "Ishinshi no kakinaoshi." *Tokyo Asahi shinbun*, July 1, 1935, evening edition: 2.

———. *Sabakuha no ketsujin*. Tokyo: Heibonsha, 1931.

Itō Masao. *Fukuzawa Yukichi no kenkyū*. Kobe: Konan Daigaku, 1966.

——. *Shiryō shūsei Meijijin no mita Fukuzawa Yukichi*. Tokyo: Keiō Tsūshin, 1970.
Ivy, Marilyn. *Discourses of the Vanishing: Modernity, Phantasm, Japan*. Chicago: University of Chicago Press, 1995.
Iwagami Masai. *Shōgaku Kōzukeshi*. Maebashi: Kankodō, 1894.
Iwanami Kōza Nihon Tsūshi 16. Kindai 1. Tokyo: Iwanami Shoten, 1994.
Iwasaki Hideshige. *Ishin zenshi; Sakurada gikyoroku*. Tokyo: Yoshikawa kōbunkan, 1911.
Jansen, Marius B. *Sakamoto Ryōma and the Meiji Restoration*. New York: Columbia University Press, 1994.
Jansen, Robert S. "Resurrection and Appropriation: Reputational Trajectories, Memory Work, and the Political Use of Historical Figures." *American Journal of Sociology* 112, no. 4 (2007): 953–1007.
Jūbishi Yoshihiko. *Oguri Kōzukenosuke no shi*. Tokyo: Daiichi Shuppansha, 1929.
Kaionji Chōgorō. *Kaionji Chōgorō zenshū*. Tokyo: Asahi Shinbunsha, 1969.
——. *Oguri Kozukenosuke*. Tokyo: Kokubunsha, 1942.
Kanakura Masami. *Furansu kōshi Rosesu to Oguri Kōzukenosuke*. Tokyo: Daiyamondo Shuppan, 1935.
Kaneko Yōbun. "Hana no shōgai wo miru: Shinbashi kabukijo dai ichibu." *Kabukikai* 11, no. 12 (1953): 61–63.
Karatani Kōjin. *Origins of Modern Japanese Literature*. Durham: Duke University Press, 1993.
Karlin, Jason G. "The Tricentennial Celebration of Tokyo: Inventing the Modern Memory of Edo." In *Image and Identity: Rethinking Japanese Cultural History*, edited by Yamaji Hidetoshi and Jeffrey E. Hanes, 215–27. Kobe: Research Institute for Economics & Business Administration, Kobe University, 2004.
Karube Tadashi. *Maruyama Masao and the Fate of Liberalism in Twentieth-Century Japan*. Tokyo: International House of Japan, 2008.
Kasahara Kazuo. *Gendai ni ikiru Nihonshi no gunzō*. Tokyo: Asoka Shuppansha, 1966.
Kashima Shigeru. "San-shimon shugisha Shibusawa Eiichi: Tokugawa Yoshinobu Kode no henshū." *Shokun* 35, no. 7 (2003): 278–84.
Katsu Kaishū. *Hikawa seiwa*. Tokyo: Kodansha, 2000.
——. *Kaishū zadan*. Tokyo: Iwanami Shoten, 1995.
Kawaguchi Sunao. *Tokugawa maizōkin kenshō jiten*. Tokyo: Shin jinbutsu Ōraisha, 2001.
Kawamura Masaru. *Hatamoto chigyōsho no shihai kōzō: Hatamoto Ishiko-shi no chigyōsho shihai to kasei kaikaku*. Tokyo: Yoshikawa Kōbunkan, 1991.
Kawasaki Shizan. *Boshin senshi*. Tokyo: Hakubunkan, 1894.
Kazuo Tsukakoshi. "Bakufu suibōron wo yomu." *Musashino Joshi Daigaku Tanki Daigakubu kiyō*, no. 1 (2000): 53–65.
Keene, Donald. *Five Modern Japanese Novelists*. New York: Columbia University Press, 2003.
Kei Emu. "Omoide no mama," *Kyūbakufu* 4, no. 4 (1901): 52–54.
Kikegawa Hiromasa. "Gaikōkan to shite no Oguri Tadamasa: 1861 nen rōkan 'Posadonikku' Gō Tsushima Taihaku Jiken." *Seiji Keizai Shigaku* 277 (1989): 15–42.
Kimura Kaishū, and Nihon Shiseki Kyokai. *Sanjūnenshi*. Tokyo: Tokyo Daigaku Shuppankai, 1978 (1892).

Kimura Ki, and Meiji Bunka Kenkyukai. *Bakumatsu Meiji shinbun zenshū*. Tokyo: Seikai Bunko, 1961.

Kimura Naomi. *Tengai no bushi: Bakushin Oguri Kōzukenosuke 1*. Tokyo: Riidosha, 2006.

———. *Tengai no bushi: Bakushin Oguri Kōzukenosuke 2*. Tokyo: Riidosha, 2006.

———. *Tengai no bushi: Bakushin Oguri Kōzukenosuke 3*. Tokyo: Riidosha, 2008.

———. *Tengai no bushi: Bakushin Oguri Kōzukenosuke 4*. Tokyo: Riidosha, 2008.

Kitajima Masamoto, ed. *Edo bakufu: Sono jitsuryokushatachi*. Tokyo: Jinbutsu Ōraisha, 1964.

Kiya Takayasu. *Bakushin Oguri Kōzukenosuke: Maizōkin yueni tsuminakushtie kiraru*. Tokyo: Tairyūsha, 1982.

Kodama Kota, Nishigaki Seiji, Yamamoto Takashi, and Yuko Ushiki, eds. *Gunmaken no rekishi*. Tokyo: Yamakawa Shuppansha, 1997.

Kodama Yoshio. *Sugamo Diary*. Tokyo, 1960.

Koitabashi Ryōhei. "Gonda no nanushi Satō Tōshichi to tōkai nikki." *Tatsunami*, no. 1 (1976): 43–48.

———. "Oguri kankei no ayamari wo kenshō suru." *Tatsunami*, no. 22 (1997): 11–13.

Koizumi Takashi. "Kaisetsu." In *Meiji jūnen teichū kōron, yasegaman no setsu*. Tokyo: Kodansha, 1985.

———. *Katsu Kaishū no raibaru Oguri Kōzukenosuke ichizoku no higeki: Oguri sōdō oyobi Oguri fujin tō dasshutsu senkō, Aizu he no michi tōsa jitsuroku*. Takasaki: Asaosha, 1999.

———. "Oguri no maizōkin no shingi wo tadasu." *Tatsunami*, no. 25 (2000): 10–14.

Kōtō gakkō Nihonshi A. Tokyo: Shimizu Shoin, 2003.

Kōtō gakkō Nihonshi A kaiteiban shidō to kenkyū. Tokyo: Shimizu Shoin, 2007.

Kurabuchi Sonshi Henshu Iinkai. *Kurabuchi sonshi*. Gunma-ken Gunma-gun Kurabuchi-mura: Kurabuchi-mura Yakuba, 1975.

Kurimoto Joun. *Hōan jisshū*. Tokyo: Hōchisha, 1892.

Kuwabara Masato, and Tanaka Akira, eds. *Hirano Yajuro bakumatsu, ishin nikki*. Sapporo-shi: Hokkaido Daigaku Tosho Kankōkai, 2000.

Kuwata Tadachika. *Nihon takarajima tanken*. Tokyo: Nihon Bungeisha, 1976.

Lebra, Joyce C. "Yano Fumio: Meiji Intellectual, Party Leader, and Bureaucrat." *Monumenta Nipponica* 20, no. 1/2 (1965): 1–14.

Lebra, Takie Sugiyama. *Above the Clouds: Status Culture of the Modern Japanese Nobility*. Berkeley: University of California Press, 1993.

Lebra-Chapman, Joyce. *Okuma Shigenobu: Statesman of Meiji Japan*. Canberra: Australian National University Press, 1973.

Legg, S. "Sites of Counter-Memory: The Refusal to Forget and the Nationalist Struggle in Colonial Delhi." *Historical Geography* 33 (2005): 180–201.

Leheny, David Richard. *The Rules of Play: National Identity and the Shaping of Japanese Leisure*. Ithaca: Cornell University Press, 2003.

———. *Think Global, Fear Local: Sex, Violence, and Anxiety in Contemporary Japan*. Ithaca, NY: Cornell University Press, 2006.

Lehmann, Jean-Pierre. "Léon Roches—Diplomat Extraordinary in the Bakumatsu Era: An Assessment of His Personality and Policy." *Modern Asian Studies* 14, no. 2 (1980): 273–307.

Liman, Anthony V. *Ibuse Masuji: A Century Remembered.* Prague: Charles University in Prague, Karolinum Press, 2008.

Lipsitz, George. *Time Passages: Collective Memory and American Popular Culture.* Minneapolis: University of Minnesota Press, 2001.

Maehara Masami. "Media sangyō to kankō sangyō—taiga durama to kankō bijinesu." *Tōyōgakuen Daigaku Kiyō* (2008): 313–150.

Maruyama Masao. "Ogyū Sōrai no zōi mondai." In *Kindai Nihon on kokka to shisō.* Edited by Ienaga Saburō kyoju Tokyo kyoiku daigaku taikan kinen ronshū kankō ii'in kai. Tokyo: Sanseido, 1979.

Marcus, Marvin. *Paragons of the Ordinary: The Biographical Literature of Mori Ōgai.* Honolulu: University of Hawai'i Press, 1993.

Mashimo Kikugoro. *Yanada sensekishi: Meiji Boshin.* Gunma-ken Ora-gun Koizumi-cho: Yanada Sensekishi Hensan Kōenkai, 1923.

Matsumoto Sannosuke. In Fukuzawa Yukichi and Eiichi Kiyooka, *The Autobiography of Yukichi Fukuzawa.* New York: Columbia University Press, 2007.

Matsuo Masahito. *Meiji Ishin to bunmei kaika.* Tokyo: Yoshikawa Kōbunkan, 2004.

———. *Nihon no jidaishi.* Tokyo: Yoshikawa Kōbunkan, 2004.

Matsuoka Masao. *Sōmu iinkai 8* (159). March 30, 2004.

McClain, James L. *Edo and Paris: Urban Life and the State in the Early Modern Era.* Ithaca, NY: Cornell University Press, 1994.

———. *Japan: A Modern History.* New York: W. W. Norton & Company, 2002.

McCormack, Gavan. *The Emptiness of Japanese Affluence.* New York: M. E. Sharpe, 2001.

McKeown, Elizabeth. "Local Memories." *U.S. Catholic Historian* 21, no. 2 (2003): 19–29.

Medzini, Meron. *French Policy in Japan during the Closing Years of the Tokugawa Regime.* Harvard East Asian Monographs 41. Cambridge: Harvard University Press, 1971.

Mehl, Margaret. *History and the State in Nineteenth-Century Japan.* New York: St. Martin's Press, 1998.

"Meiji no chichi: Oguri Kōzukenosuke." *Shūkan New York seikatsu,* January 1, 2010.

Mertz, John Pierre. *Novel Japan: Spaces of Nationhood in Early Meiji Narrative, 1870–88.* Michigan Monograph Series in Japanese Studies 48. Ann Arbor, MI: Center for Japanese Studies, University of Michigan, 2003.

Minami Kazuo. *Bakumatsu Edo shakai no kenkyū.* Tokyo: Yoshikawa Kōbunkan, 1978.

Mino Yukinori. "Kindai ikōki, kanryō soshiki hensei ni okeru bakufu kanryō ni kansuru tōkeiteki kento." In *Kinsei kokka no kenryoku kōzō: seiji, shihai, gyōsei,* edited by Ōishi Manabu, 379–419. Tokyo: Iwata Shoin, 2003.

Minomura Seiichirō. *Minomura Rizaemon den.* 4th ed. Tokyo: Minomura Gōmei kaisha, 1987.

Mishima Chūfū. "Mishima Chūshū kō danwa." *Kyūbakufu* 2 no. 9 (1900): 85–87.

Mitsui ginkō hachijūnenshi hensan iinkai. *Mitsui ginko hachijūnenshi.* Tokyo: Mitsui Ginkō, 1957.

Miyazawa Seiichi. "Bakumatsu ishin he no kaiki (2) henbō 'kakumei' no romantizumu." *Kyūshū kokusai daigaku kyōyō kenkyū* 9, no. 2:24 (2002): 1–22.

———. *Meiji ishin no saisōzō: kindai Nihon no "kigen shinwa."* Tokyo: Aoki Shoten, 2005.

Miyoshi Masao. *As We Saw Them: The First Japanese Embassy to the United States (1860).* Berkeley: University of California Press, 1979.

Mizuno Tomoyuki. *Akagi ōgon tsuiseki: Mizunoke sandai, noshūnen maizōkin hakkutsu 100 nen.* Tokyo: Magajinhausu, 1994.

Monnet, Livia. *Approches critiques de la pensée Japonaise du XXe siècle.* Montreal: Les presses de l'université de Montreal, 2001.

Morimura Munefuyu. *Yoshitsune densetsu to Nihonjin.* Tokyo: Heibonsha, 2005.

Moriki Akira. "Meiji Ishin rejiimu kara no dakkyaku." *Jiyū* 50 (5), 2008: 55–64.

Morita Akio, and Shintaro Ishihara. *"No" to ieru Nihon: Shin Nichi-Bei kankei no kado.* Tokyo: Kōbunsha, 1989.

Morris, Ivan I. *The Nobility of Failure: Tragic Heroes in the History of Japan.* Holt, Rinehart and Winston, 1975.

Morris, John. "Hatamoto Rule: A Study of the Tokugawa Polity as a Seigneurial System." *Papers on Far Eastern History,* no. 41 (1990): 9–44.

Morris-Suzuki, Tessa. *The Past within Us: Media, Memory, History.* New York: Verso, 2005.

Morse, Ronald A. *Yanagita Kunio and the Folklore Movement: The Search for Japan's National Character and Distinctiveness.* New York: Garland Publishing, 1990.

Motoyama Yukihiko. *Proliferating Talent: Essays on Politics, Thought, and Education in the Meiji Era.* Honolulu: University of Hawai'i Press, 1997.

Murakami Shōken. "Jōshū kō no ihin ni tsuite." *Tatsunami,* no. 2 (1977): 12–16.

Murakami Taiken. *Nanushi Satō Tōshichi no sekai isshū: Oguri Tadamasa jusha no Kiroku.* Maebashi: Jōmō Shinbunsha, 2001.

———. "Oguri Kōzukenosuke kankei shoseki annai. http://www5.wind.ne.jp/cgv/oguri/book.htm.

———, ed. *Oguri Tadamasa no subete.* Tokyo: Shin Jinbutsu Ōraisha, 2008.

Murakami Taiken, and Satō Tōshichi. *Bakumatsu kenbei shisetsu Oguri Tadamasa jūsha no kiroku: Nanushi Satō Tōshichi no sekai isshū.* Maebashi: Hatsubai Jōmō Shinbunsha Shuppankyoku, 2001.

Murata Ujihisa, and Sasaki Chihiro. *Zoku saimu kiji.* Tokyo: Tokyo Daigaku Shuppankai, 1974.

Nagai Mai. "Meijiki ni okeru kyūbakushin to sabakuha shikan." *Nihonshi no hōhō,* no. 6 (2007): 45–53.

Naimusho shohitsu, no. 93. Japanese National Diet Archives.

Najita Tetsuo, and Joint Committee on Japanese Studies. *Conflict in Modern Japanese History: The Neglected Tradition.* Princeton, NJ: Princeton University Press, 1982.

Nakajima Akira. *Jōshū no Meiji ishin.* Maebashi: Miyama Bunko, 1996.

Nakajima Mineo. *Bakushin Fukuzawa Yukichi.* Tokyo: TBS Buritanika, 1991.

Nakamura Kaoru. *Kanda bunkashi.* Tokyo: Kanda Shiseki Kenkyūkai, 1935.

Nakamura Katsumaro, and Akimoto Shunkichi. *Lord Ii Naosuke and New Japan.* Yokohama: Japan Times, 1909.

Nakamura Kazuo, and Kishō Sasaki. "Shogatsu jidaigeki 'Matamo yametaka teishu dono' no seisaku: Bakumatsu no meibugyō Oguri Kōzukenosuke." *Eiga Terebi Gijutsu* 605 (2001): 24–26.

Nakamura Yasuhiro. *Satō Issai, Asaka Gonsai*. Tokyo: Meitoku Shuppansha, 2008.

Nakano Kōji. *Seihin no shisō*. Tokyo: Sōshisha, 1992.

Nakazato Kaizan. *Daibosatsu Tōge: Great Bodhisattva Pass*. Tokyo: Shunjusha, 1929.

Narita Ryūichi. "Historical Practice before the Dawn: 'Modern Japan' in Postwar History." *Iichiko Intercultural* 7 (1995).

———. *"Kokyō" to iu monogatari: Toshi kūkan no rekishigaku*. Tokyo: Yoshikawa Kōbunkan, 1998.

———. *"Sensō keiken" no sengoshi: Katarareta taiken/shōgen/kioku*. Tokyo: Iwanami Shoten, 2010.

———. *Shiba Ryōtarō no bakumatsu Meiji: "Ryōma ga yuku" to "Saka no ue no kumo" wo yomu*. Tokyo: Asahi Shinbunsha, 2003.

Narumi Fū. *Dotō sakamakumo: Nihon kindaika o michibiita Ono Tomogorō to Oguri Tadamasa*. Tokyo: Shin Jinbutsu Ōraisha, 2009.

Narushima Motonao, and Kuroita Katsumi. *Tokugawa jikki*. Tokyo: Yoshikawa Kōbunkan, 1964.

National Diet Archives, Ondai 9, 105, October 20, 1928.

Neary, Ian, ed. *Leaders and Leadership in Japan*. United Kingdom: Curzon Press, 1996.

Nihon Kanji Nōryuku Kentei Kyōkai. "2007 nen kotoshi no kanji," http://www.kanken .or.jp/kanji/kanji2007/kanji.html.

Ninagawa Arata. *Ajia ni ikiru no michi*. Tokyo: Nihon Shoin, 1929.

———. *Ishin seikan*. Tokyo: Chiyoda Shoin, 1952.

———. *Ishin zengo no seisō to Oguri Kōzuke no shi*. Tokyo: Nihon Shoin, 1928.

———. *Ishin zengo no seisō to Oguri Kōzuke: Zoku*. Tokyo: Nihon Shoin Shuppanbu, 1931.

———. *Kaikoku no senkakusha: Oguri Kōzuke no Suke*. Tokyo: Chiyoda Shoin, 1953.

———. *Kōshūwan no senryō to Karafuto no senryō*. Tokyo: Shimizu Shoten, 1914.

———. *Les Réclamations Japonaises et le droit international*. Paris: Pedone, 1919.

———. *Tennō: Dare ga Nihon minzoku no shujin de aru ka*. Tokyo: Kōbunsha, 1952.

Ninagawa Arata, and Abe Dōzan. "Oguri Kōzukenosuke to Fumon'in." *Saitama Shidan* 8, no. 1:13–22.

Nishikata Kyoko. *Jōmō karuta no kokoro: Urano Masahiko no hansei*. Maebashi: Gunma Bunka Kyōkai, 2002.

Nishiwaki Yasushi. *Hatamoto Mishima Masakiyo nikki: Bakumatsu ishinki wo ikita hatamoto mizukara no kiroku*. Tokyo: Tokugawashi Hatamoto Fujitsuki Mishima Shi Yonhyakunenshi Kankōkai, 1987.

Noguchi Takehiko. *Edo wa moeteiruka*. Tokyo: Bungei Shunjō, 2006.

———. "Karakkaze Akagisan: Oguri Kōzukenosuke no saiki." *Bungakukai* 58, no. 8 (2004): 167–89.

Nolletti, Arthur. *Reframing Japanese Cinema: Authorship, Genre, History*. Bloomington: Indiana University Press, 1992.

Numata Jiro. "Shigeno Yasutsugu and the Modern Tokyo Tradition of Historical Writing." In *Historians of China and Japan*, edited by W. G. Beasley and F. G. Pulleyblank. London: Oxford University Press, 1961.

Nume Hiroshi, and Shōzō Shibata. "'Boshin Sensō' 130nen sadankai." *Hoppō Fudo* 37 (1999): 78–93.

Ochiai Nobutaka. *Nihonshi liburetto: Hasshū mawari to bakuto*. Tokyo: Yamakawa Shuppansha, 2002.

Ogawa Kyōichi, ed. *Kanseifu ikō hatamotoke hyakka jiten.* Tokyo: Tōyō Shorin, 1997–98.

Oguri Kuniko. "Oguri Kōzukenosuke no mago kara." *Yomiuri shinbun,* September 16, 1956.

Oguri Mataichi. *Ryūkei Yano Fumio-kun den: Denki Yano Fumio.* Tokyo: Ozorasha, 1993.

Oguri Tadahito. "Nichirō sensō to Oguri Tadamasa." *Tatsunami,* no. 14 (1989): 11–15.

———. "Ninagawa hakase to sono meicho." *Tatsunami,* no. 6 (1981): 1–8.

Oguri Tadamasa. *Oguri nikki. Gunma-ken shiryōshū* 7:5–87.

Ōishi Manabu. *Kinsei kokka no kenryoku kōzō: seiji, shihai, gyōsei.* Tokyo: Iwata Shoin, 2003.

Oka Shigeki. *Ii Tairō.* Tokyo: Sawamoto Shobō, 1948.

Okada Shōji. "Gunma-ken ni okeru shishi hensan jigyō to sono hensen." *Sōbun,* no. 24 (2007): 1–25.

Ōkubo Toshiaki. *Nihon kindai shigaku no seiritsu.* Tokyo: Yoshikawa Kōbunkan, 1988.

———. *Sabakuha no rongi.* Tokyo: Yoshikawa Kōbunkan, 1986.

Ōkuma Shigenobu. *Fifty Years of New Japan: Volume One.* London: Smith Elder, 1909.

Olick, Jeffrey K. *States of Memory: Continuities, Conflicts, and Transformations in National Retrospection.* Durham: Duke University Press, 2003.

Ōmiya-shi. *Ōmiyashi shi.* Ōmiya-shi: Ōmiya Shiyakusho, 1968.

Ooms, Herman. *Tokugawa Ideology: Early Constructs, 1570–1680.* Princeton, NJ: Princeton University Press, 1985.

Ōshima Masahiro. *Tsumi nakushite kiraru: Oguri Kōzukenosuke.* Tokyo: Shinchōsha, 1994.

Ōsone Tatsuo, director. *Ōedo no kane: Dai Tokyo tanjō.* Japan: Shochiku, 1958.

———. *Ōedo no kane: Dai Tokyo tanjo / Kishimoto, Gin'ichi, Production.* Japan: Shochiku, 1958.

Ōtashi-shi tsūshihen kinsei. Ōtashi, 1992.

Ōtsubo Shihō, and Hozumi Miharu. *Oguri Kōzukenosuke.* Yokosuka: Oguri Kōzukenosuke Shinobukai, 1975.

———. *Oguri Kōzukenosuke kenkyū shiryō ochibohiroi.* Kurabuchi: Oguri Kenshōkai, 1957.

Ozaki Hotsuki. "Kaisetsu" in *Tanaka Hidemitsu zenshū.* Vol. 8. Tokyo: Hoga Shoten, 1964.

Peterson, Brian J. "History, Memory and the Legacy of Samori in Southern Mali, c. 1880–1898." *Journal of African History* 49, no. 2 (2008): 261–79.

Pierson, John D. *Tokutomi Sohō, 1863–1957: A Journalist for Modern Japan.* Princeton, NJ: Princeton University Press, 1980.

Platt, Brian. *Burning and Building: Schooling and State Formation in Japan, 1750–1890.* Harvard East Asian Monographs 237. Cambridge: Harvard University Press, 2004.

Pyle, Kenneth B. *The New Generation in Meiji Japan: Problems of Cultural Identity, 1885–1895.* Stanford, CA: Stanford University Press, 1969.

Ravina, Mark J. "The Apocryphal Suicide of Saigō Takamori: Samurai, 'Seppuku,' and the Politics of Legend." *Journal of Asian Studies* 69, no. 3 (2010): 691–721.

———. *The Last Samurai*. New Jersey: John Wiley & Sons, 2004.

Rekishigaku Kenkyukai, ed. *Meiji ishinshi kenkyū kōza*. 7 vols. Tokyo: Heibon-sha, 1958–59.

Rieger, Bernhard. "Memory and Normality." *History and Theory* 47, no. 4 (2008): 560–72.

Rinbara Sumio. "Shōwa shoki no bakumatsu 'monogatari': Shimozawa Kan Shinsengumi shimatsuki no shūhen." *Kobe daigaku bungakubu kiyō*, no. 27 (2000): 453–68.

Roberts, John G. *Mitsui Empire: Three Centuries of Japanese Business*. New York: Weatherhill, 1973.

Robertson, Jennifer. "Les 'Bataillons fertiles': Sexe et la citoyenneté dans le Japon Impérial." In *New Critical Approaches to Twentieth-Century Japanese Thought*, edited by Livia Monnet, 275–301. Montreal: University of Montreal Press.

———. *Native and Newcomer: Making and Remaking a Japanese City*. Berkeley: University of California Press, 1991.

Rousso, Henry. Translated by Arthur Goldhammer. *The Vichy Syndrome: History and Memory in France since 1944*. Cambridge: Harvard University Press, 1994.

Saaler, Sven, and Wolfgang Schwentker. *The Power of Memory in Modern Japan*. Folkestone, UK: Global Oriental, 2008.

Saitō Takashi. "Sengo jidai shōsetsu no shisō: mo hitotsu no 1960 nendairon." *Shisō no kagaku*, no. 67 (1976): 11–22.

Sakamoto Fujiyoshi. *Bakumatsu ishin no keizaijin: senkenryoku, ketsudanryoku, shidōryoku*. Tokyo: Chūō Kōronsha, 1984.

———. *Oguri Kōzukenosuke no shōgai: Hyōgo Shosha wo tsukutta saigo no bakushin*. Tokyo: Kodansha, 1987.

Sakaya Taichi. "Sangiin kokumin seikatsu-keizai ni kansuru chōsakai." National Diet Archive. January 28, 2009.

Sanō Shishi Hensan Iinkai. *Sanō-shishi, tsūshihen*. Sanō: Sanō-shi, 1978.

Sasaki Suguru. "Saigō Takamori to Saigō densetsu." In *Iwanami kōza Nihon tsūshi 16: Kindai 1*. Edited by Asao Naohiro, Amino Yoshihiko, Ishii Susumu, Kano Masanao, Hayase Shōhachi, and Yasumaru Yoshio, 325–340. Tokyo: Iwanami Shoten, 1994.

Satō Hisao. "Watashi no ie ni tsutawaru Oguri Kōzukenosuke kō no hanashi." *Tatsunami*, no. 4 (1979): 22–25.

Satō Masami. *Kakugo no hito: Oguri Kōzukenosuke Tadamasa den*. Tokyo: Kadokawa Shoten, 2009.

Satō, Tadao. *Iji no bigaku: Jidaigeki eiga taizen*. Tokyo: Jakometei Shuppan, 2009.

Satoh, H. (Henry) and Shimada Saburō. *Agitated Japan: The Life of Baron Ii Kamon-No-Kami Naosuke (Based on the Kaikoku shimatsu of Shimada Saburō)*. Tokyo: Dai Nippon Tosho; New York: D. Appleton, 1896.

Schwartz, Barry. *Abraham Lincoln and the Forge of National Memory*. Chicago: University of Chicago Press, 2000.

Schwartz, Barry, and Howard Schuman. "History, Commemoration, and Belief: Abraham Lincoln in American Memory, 1945–2001." *American Sociological Review* 70, no. 2 (2005): 183–203.

Seta Tōyō. *Shōnen dokuhon #40: Oguri Kōzukenosuke*. Tokyo: Hakubunkan, 1901.
Sheldon, Charles D. "Scapegoat or Instigator of Japanese Aggression? Inoue Kiyo-shi's Case Against the Emperor." *Modern Asian Studies* 12, no. 1 (1978): 1–35.
Shiba Ryōtarō. *Jūichibanme no shishi*. Tōkyō: Bungei Shunjū, 1967.
———. *The Last Shogun: The Life of Tokugawa Yoshinobu*. New York: Kodansha International, 1998.
———. *Meiji to iu kokka*. Tokyo: Nihon Hōsō Shuppan Kyōkai, 1989.
———. *Moeyoken*. Vol. 1. Tokyo: Shinchōsha, 1977.
———. *Moeyoken*. Vol. 2. Tokyo: Shinchōsha, 1978.
———. *Shiba Ryōtarō rekishi kandan*. Tokyo: Chūō Kōron Shinsha, 2000.
———. *Shiba Ryōtarō zenshu*. Tokyo, Bungei Shunjū, 1971.
"Shibukawashi Akagi shōkōkai: ōgon no nemuru mura: Mizunoke no hakkutsu jigyō." http://www7.wind.ne.jp/akagi-shoko/maizo/maizo02.htm.
Shibusawa Eiichi and Teruko Craig. *The Autobiography of Shibusawa Eiichi: From Peasant to Entrepreneur*. Tokyo: University of Tokyo Press, 1994.
Shidankai. *Senbō junnan shishi jinmeiroku: kaiei gannen yori Meiji nijūsannen ni itoru no kikan*. Tokyo: Hara Shobō, 1907.
Shimada Saburō. *Dōhōkai hōkoku 1* (1). Tokyo Rittaisha, 1896.
———. *Kaikoku shimatsu: Ii Kamon no kami Naosuke den*. Tokyo: Yoronsha, 1888.
Shimane Kiyoshi. *Tenkō: Meiji ishin to bakushin*. Tokyo: Sanichi Shobō, 1969.
Shimazu Naoko. *Nationalisms in Japan*. Hoboken: Taylor & Francis, 2006.
———. "Popular Representations of the Past: The Case of Postwar Japan." *Journal of Contemporary History* 38, no. 1 (2003): 101–16.
Shimoda Hiraku. "Between Homeland and Nation: Aizu in Early and Modern Japan," PhD diss., Harvard University, 2005.
"Shinbun," *Yomiuri shinbun*, November 24, 1880.
Shinchosha. *Rekishi shōsetsu no seiki*. Shincho bunko. Tokyo: Shinchosha, 2000.
Shirayanagi Natsuo. "Oguri Kōzukenosuke ibun." *Senshū Daigaku Gakkai* 43 (1987): 1–57.
"Shisetsu no goannai—Hamayū Sansō." http://www6.wind.ne.jp/hamayu/sisetu/sisetu.html.
Shizuoka-ken. *Shizuoka kenshi. Shiryōhen 16 kingendai 1*. Shizuoka-shi: Shizuoka-ken, 1989.
Sims, Richard. *French Policy Towards the Bakufu and Meiji Japan, 1854–95*. Surrey: Curzon Press, 1998.
Sippel, Patricia. "Popular Protest in Early Modern Japan: The Bushū Outburst." *Harvard Journal of Asiatic Studies* 37, no. 2 (1977): 273–322.
Steele, William. "Against the Restoration: Katsu Kaishū's Attempt to Reinstate the Tokugawa Family." *Monumenta Nipponica* 36, no. 3 (1981): 299–316.
———. *Alternative Narratives in Modern Japanese History*. New York: Routledge Curzon, 2003.
———. "Edo in 1868: The View from Below." *Monumenta Nipponica* 45, no. 2 (1990): 127–55.
———. "Katsu Kaishū and the Collapse of the Tokugawa Bakufu." PhD diss., Harvard, 1976.
———. "*Yasegaman no setsu*: On Fighting to the Bitter End." *Asian Cultural Studies Special Issue* 11 (2002): 139–52.

Stegewerns, Dick. *Nationalism and Internationalism in Imperial Japan: Autonomy, Asian Brotherhood, or World Citizenship?* London and New York: Routledge Curzon, 2003.

Suda Tsutomu. *"Akuto" no jūkyūseiki: Minshū undo no henshitsu to "kindai ikoki."* Tokyo: Aoki Shoten, 2002.

Suga Yasuo. "Shinkokugeki no 'Ii tairō.'" *Engeki hyōron* "Shinkokugeki" 1, no. 2 (1953): 50–52.

Sugiyama Hirokazu. "Hito hatamoto no Meiji ishin: Hatamoto mishima masakiyo no nikki wo Sozai to shite." *Ajia Bunkashi Kenkyū* 10 (March 2010): 1–23.

Suzuki Haruzō. "Fukuchi Ouchi no rekishikan ni tsuite." *Rekishi Kenkyū* 7–8 (1969).

Swale, Alistair. "Tokutomi Sohō and the Problem of the Nation-State in an Imperialist World." In *Nationalism and Internationalism in Imperial Japan: Autonomy, Asian Brotherhood, or World Citizenship?* edited by Dick Stegewerns, 69–88. London and New York: Routledge Curzon, 2003.

Tai, Hue-Tam Ho. "Remembered Realms: Pierre Nora and French National Memory." *American Historical Review* 106, no. 3 (2001): 906–22.

Tajiri Tasuku. *Zōi shoken den*. Tokyo: Kokuyūsha, 1927.

Takada Yūsuke. "Ishin no kioku to 'kinnō shishi' sōshutsu: Tanaka Mitsuaki no kenshō katsudō wo chūshin ni." *Hisutoria* 204 (March 2007).

Takahashi Kyōichi. *Yokosuka zōsenjo zōsetsu to sono futari onjin*. Yokosuka: Yokosuka Shiyakusho, 1952.

Takahashi Shutei. *Kinsei Jōmō ijin den: Zen*. Tokyo: Agatsuma Shokan, 1982.

Takahashi Yoshio. *Nippon taihen: Oguri Kōzukenosuke to Minomura Rizaemon*. Shueisha bunko. Tokyo: Shueisha, 1999.

Takahori Fuyuhiko. "'Hisaichi Fukushima' kara 'Yae to Kakuma no Fukushima' he: shichōritsu no yoshiakashi dake dewa hakarenai Taiga dorama no eikyōroku." *Gendai Bijinesu*. February 27, 2013: 1–3. http://gendai.ismedia.jp/articles/-/34985.

Takai Kōzanden Hensan Iinkai. *Takai Kōzan den*. Nagano-ken Kamitakai-gun Obuse-machi: Obuse-machi, 1988.

Takaki Hiroshi. "'Kyōdoai' to 'aikokushin' wo tsunagumono: Kindai ni okeru 'kyūhan' no kenshō." *Rekishi Hyōron* 659 (March 2005): 2–18.

Takasakishi. "Oguri no sato seibi kihon keikaku no sakutei." http://www.city.takasaki.gunma.jp/shisho/kurabuchi/chiiki/oguri.htm.

———. *Shinpen Kurabuchi sonshi.1 shiryōhen 1, genshi kodai chūsei kinsei*. Takasaki: Kurabuchi Sonshi Kankō Iinkai, 2008.

———. *Shinpen Kurabuchi sonshi.2 shiryōhen 2, kindai gendai*. Takasaki: Kurabuchi Sonshi Kankō Iinkai, 2008.

———. *Shinpen Takasaki shishi. shiryōshu 5, kinsei 1*, 2002.

———. *Takasaki shishi 5*, 1993.

Tamamuro Taijo. *Saigō Takamori*. Tokyo: Iwanami Shoten, 1960.

Tamura Eitarō. "Jōshū yonaoshi to Oguri Kōzukenosuke." *Yuibutsuron Kenkyū* 20 (1934): 94–109.

———. *Katsu Rintarō*. Tokyo: Yūzankaku, 1967.

———. *Yonaoshi*. Tokyo: Yūzankaku, 1960.

Tanaka Hidemitsu. "Sakurada no yuki." *Tanaka Hidemitsu zenshū*. Vol. 8. Tokyo: Hoga Shoten, 1964.

Tanaka Satoru. *Aizu to iu shinwa: "Futatsu no sengo" wo meguru "shisha no sei-jigaku."* Kyoto: Mineruva Shobō, 2010.

Tanizaki Junichirō. *Tanizaki Junichirō zenshū.* Tokyo: Chūō Kōronsha, 1981.

Taylor, Alan. "The Early Republic's Supernatural Economy: Treasure Seeking in the American Northeast, 1780–1830." *American Quarterly* 38, no. 1 (1986): 6–34.

Tenkyū Gorō. *Oguri Kōzukenosuke no hihō.* Tokyo: Shinjumbutsu Ōraisha, 1991.

Thornton, Sybil Anne. *The Japanese Period Film: A Critical Analysis.* Jefferson, NC: McFarland & Co., 2008.

Togawa Zanka. "Oguri Kōzukenosuke." *Kyūbakufu* 4, no. 8 (1900): 31–37.

———. "Shiden Oguri Kōzukenosuke." *Kyūbakufu* 4, no. 7 (1900): 29–35.

Tokita Tōkō. *Ze ya hi ya Ii Tairō.* Tokyo: Seizando, 1911.

Tokuda Atsushi. "The Origin of the Corporation in Meiji Japan." http://www.cefims .ac.uk/documents/research-9.pdf.

"Tokugawa maizōkin 'ana hori asobi' ni yonoku en shusshi shite kaisha shachō." *Shūkan bunshū*, March 3, 2000.

Tokugawa Yoshinobu. *Sekimukai hikki: Tokugawa Yoshinobu kō kaisōdan.* Tokyo: Heibonsha, 1966.

Tomita Hitoshi, and Akira Nishibori. *Yokosuka seitetsujo no hitobito: hanahiraku fransu bunka.* Yokohama: Yūrindō, 1983.

Tompkins, Tom. *Yokosuka: Base of an Empire.* Novato, CA: Presidio Press, 1981.

Torao Tatsuya. "Zōi no shopoteki kōsatsu." *Nihon Rekishi*, no. 521 (October 1991).

Totman, Conrad D. *The Collapse of the Tokugawa Bakufu, 1862–1868.* Honolulu: University of Hawai'i Press, 1980.

Touchet, Elisabeth de. *Quand les Francais armaient le Japon: La creation de l'arsenal de Yokosuka 1865–1882.* Rennes: Presses Universitaires de Rennes, 2003.

Toyokuni Kakudō. "Oguri Jōshū no shukyū no shozaichi nit suite." *Jōmō oyobi Jōmōjin* 220 (1935): 61–62.

———. "Yokosuka kaikō gojūnen shukuten ni saishite, Oguri Kōzukenosuke no tame ni hitokoto su." *Jōmō oyobi Jōmōjin* 7 (1917): 24–30.

"Tozenji." http://tozenzi.cside.com/.

Treat, John Whittier. *Pools of Water, Pillars of Fire: The Literature of Ibuse Masuji.* Seattle: University of Washington Press, 1988.

Troyansky, David G. "Memorializing Saint-Quentin: Monuments, Inaugurations and History in the Third Republic." *French History* 13, no. 1 (1999): 48–76.

Tsukagoshi Teishun (Yoshitarō). *Oguri Kōzukenosuke matsuro jiseki*, 1915.

———. "Shiron: Oguri Jōshū." *Kokumin no tomo*, March 13, 1893: 442–43.

Tsunabuchi Kenjō. *Bakushin retsuden.* Tokyo: Chuo Kōronsha, 1984.

Tsutsui Kiyotada. *Jidaigeki eiga no shisō: Nosutaruji no yukue.* Tokyo: PHP Ken-kyujo, 2000.

Tsuzukidani Maki. "Shinsengumi 'fukken' he no keifu: Shiba Ryōtarō no rekishi kōchiku." *Waseda daigaku daigakuin kyōikugaku kenkyū kiyō bekkan* 17, no. 1 (2009): 11–22.

Uchino Santoku. "Dōzan no Oguri Kōzukenosuke wo yomu." *Shomotsu Tenbō* 12, no. 2 (1942): 46–50.

Uemura Midori. *Ishin no kage Oguri Kōzukensouke.* Tokyo: Bungei shobō, 2003.

Vlastos, Stephen. "Agrarianism without Tradition: The Radical Critique of Prewar Japanese Modernity." In *Mirror of Modernity: Invented Traditions of Modern*

Japan, edited by Stephen Vlastos, 79–94. Berkeley: University of California Press, 1998.

Wakabayashi Atsushi. *Hatamotoryō no kenkyū*. Tokyo: Yoshikawa Kōbunkan, 1987.

Walthall, Anne. *The Weak Body of a Useless Woman: Matsuo Taseko and the Meiji Restoration*. Chicago, IL: University of Chicago Press, 1998.

Ward, Robert Edward, and Hajime Watanabe. *Japanese Political Science: A Guide to Japanese Reference and Research Materials*. Ann Arbor: University of Michigan Press, 1961.

Watanabe Katei, and Katsu Kaishū. *Ishin genkun Saigō Takamori-kun no den: Ishin genkun*. Tokyo: Bunjido, 1889.

Waters, Neil L. *Japan's Local Pragmatists: The Transition from Bakumatsu to Meiji in the Kawasaki Region*. Harvard East Asian Monographs 105. Cambridge: Harvard University Press, 1983.

Wertsch, James V. *Voices of Collective Remembering*. Cambridge, UK; New York: Cambridge University Press, 2002.

Wigen, Kären. *A Malleable Map: Geographies of Restoration in Central Japan, 1600–1912*. Berkeley: University of California Press, 2010.

———. "Teaching about Home: Geography at Work in the Prewar Nagano Classroom." *Journal of Asian Studies* 59, no. 3 (2000): 550–74.

Wigmore, John Henry. *Law and Justice in Tokugawa Japan: Materials for the History of Japanese Law and Justice under the Tokugawa Shogunate 1603–1867*. Tokyo: University of Tokyo Press, 1967.

Wray, Harry, and Hilary Conroy. *Japan Examined: Perspectives on Modern Japanese History*. Honolulu: University of Hawai'i Press, 1983.

Yajima Hiroaki. *Oguri Kōzukenosuke Tadamasa: Sono nazo no jimbutsu no sei to shi jika 200 chō en to iwareru Tokugawa bakufu no maizōkin! Sono shinjutsu wo shiru tada hitori jimbutsu repooto*. Tokyo: Gunma Shuppan Sentaa, 1992.

Yamada Takemaro. *Gunmaken shiryōshu*. Maehashi: Gunma-ken Bunka Jigyō Shinkōkai, 1966.

Yamaguchi Masao. "*Haisha*" *no seishinshi*. Tokyo: Iwanami Shoten, 1995.

Yamaguchi Takeo, ed. *Nakanojōmachi bakumatsu no uchikowashi to Oguri Kōzukenosuke*. Nakanojō: Nakanojōmachi Kyōikuiinkai, 1985.

"Yamaguchiken Hagishi kara." *Yomiuri shinbun Fukushima*, 3 March 2011.

Yamakawa Kenjirō. *Aizu Boshin senshi*. Tokyo: Aizu Boshin Senshi Hensankai, 1933.

Yamaki Minoru. "130 nen no onshū wo koete: Chōshū, Aizu 'rekishiteki wakai' ni kasoku: Rekikenzenkoku taikai ga suishin ni hito yaku." *Rekishi Kenkyū* 451 (1998): 48–52.

Yamamura Kozo. *A Study of Samurai Income and Entrepreneurship; Quantitative Analyses of Economic and Social Aspects of the Samurai in Tokugawa and Meiji, Japan*. Harvard East Asian Series 76. Cambridge, MA: Harvard University Press, 1974.

Yamauchi Masayuki. *Bakumatsu ishin ni manabu genzai*. Tokyo: Chuo Koron Shinsha, 2010.

Yamazaki Minoru. *Takai Kōzan yumemonogatari*. Nagano: Takai Kōzan Kinenkan, 2004.

Yokosuka kaigun senshō hen. *Yokosuka kaigun senshōshi*. Tokyo: Hara Shobō, 1915.

Yokosuka kaikokushi kenkyukai. "Shinpojiumu: Oguri Kōzukenosuke." *Yokosuka kaikokushi kenkyū* 1 (2001): 32–53.

———. "Shinpojiumu: Oguri Kōzukenosuke." *Yokosuka kaikokushi kenkyū* 2 (2002): 36–44.

Yokosukashi. *Oguri Kōzukenosuke Tadamasa: Kindai Nihon no ishizue o kizuita ijin.* Yokosuka: Yokosukashi, 2000.

———. *Yokosuka annaiki.* Yokosuka: Yokosuka Kaiko Gojunen Shukugakai, 1915.

Yoneyama, Lisa. *Perilous Memories: The Asia Pacific Wars.* Durham: Duke University Press, 2001.

Yoshida Michio. *Jinpangu no fune: Oguri Kōzukenosuke kokka hyakunen no kei jōkan (gekan).* Tokyo: Kōjinsha, 2001.

Yoshiimachi. *Yoshii chōshi.* Yoshiimachi: Yoshiichoshihensan'iinkai, 1974.

Yoshioka Michio. *Jipangu no fune: Oguri Kōzukenosuke kokka hyakunen no kei.* Tokyo: Kojinsha, 2001.

Yujiro Oguchi. "The Reality behind Musui Dokugen: The World of the Hatamoto and Gokenin." Translated by Gaynor Sekimori. *Journal of Japanese Studies* 16, no. 2 (1990): 289–308.

"Yūsei mineika ni kansuru tokubetsu iinakai," no. 15, August 15, 2005. Diet Record.

Index

Harvard East Asian Monographs
(most recent titles)